How to
Save the World
for Just a
Trillion Dollars

ALSO BY ROWAN HOOPER

Superhuman: Life at the Extremes of Our Capacity

How to Save the World for Just a Trillion Dollars

The Ten Biggest Problems We Can Actually Fix

ROWAN HOOPER

THE EXPERIMENT
NEW YORK

The Experiment, LLC
220 East 23rd Street, Suite 600
New York, NY 10010-4658
theexperimentpublishing.com

THE EXPERIMENT and its colophon are registered trademarks of The Experiment, LLC. Many of the designations used by manufacturers and sellers to distinguish their products are claimed as trademarks. Where those designations appear in this book and The Experiment was aware of a trademark claim, the designations have been capitalized.

The Experiment's books are available at special discounts when purchased in bulk for premiums and sales promotions as well as for fundraising or educational use. For details, contact us at info@theexperimentpublishing.com.

Library of Congress Cataloging-in-Publication Data available upon request

ISBN 978-1-61519-828-3
Ebook ISBN 978-1-61519-829-0

Cover and text design by Beth Bugler

Manufactured in the United States of America

First printing May 2022
10 9 8 7 6 5 4 3 2 1

For my mum, Mary

Contents

Jeff Bezos, on course to become the world's first trillionaire, at an Amazon event in New Delhi, 2020

Introduction

Project Trillion

YOU KNOW THAT DAYDREAM where you suddenly come into a vast fortune? You could buy a castle or a tropical island hideaway, help out all your friends, do a bit of good in the world. But what if it was a truly incredible sum? What if you had a *trillion dollars* to spend, and a year to do it? And what if the rules of the game were that you had to do it for the world—make some real difference to people's lives, or to the health of the planet, or to the advancement of science.

A trillion dollars—that's *one thousand billion* dollars—is at once an absurdly huge amount of money, and not that much in the scheme of things. It is, give or take, 1 percent of world GDP. It's what the United States spends every year and a half on the military, or in less time if there's a big war on. It is an amount that can be quite easily rustled up through the smoke and mirrors of quantitative easing, which officially is the mass purchase of government bonds but which looks suspiciously like the spontaneous creation of money. After the 2008 financial crash, more than $4.5 trillion was quantitatively eased in the US alone.[1] All the other major economies made their own money in this ghostly way.

And it is not just governments that have this kind of money. Two of the world's biggest companies, Microsoft and Apple, are each worth over $2 trillion; Amazon isn't far behind. Amazon boss Jeff Bezos has a personal fortune of around $200 billion, and could become a trillionaire—the world's first—by 2026, while the world's richest 1 percent together own a staggering $162 trillion. That's 45 percent of all global wealth.[2]

There is so much money out there, sloshing around. At the start of 2020, private equity firms held $1.45 trillion in what they call "dry powder," and what the rest of us call "cash": piles of money sitting around awaiting investment.[3] Just *imagine* what you could do with it. Just a little bit of it, just a measly trillion dollars. You could send probes across the solar system. You could eradicate malaria— hell, you could cure *all* diseases. You could start a settlement on the Moon. You could launch an interstellar mission to another star. You could build a massive particle collider to explore the nature of reality like never before. You could solve global poverty. You could build new kinds of quantum computers and try to develop artificial consciousness. You could work to increase human life span. Then again, maybe you could try to transition the world to renewable energy. You could buy and preserve the rainforests. You could try to save all endangered species. You could refreeze the melting Arctic. You could reduce the amount of carbon dioxide (CO_2) in the atmosphere. You could launch a new, sustainable agricultural revolution. You could even create a new life-form.

If it sounds like I'm getting carried away, let me just say that all these ideas are projects that scientists are thinking about and even working on, but that are hampered by lack of resources. The world is full of extraordinary opportunities, and the vast majority are never undertaken. Those challenges that *are* tackled mostly either fail or only inch forward imperceptibly and infuriatingly slowly. And the problems of the world, most of which we've created, are so expensive to solve that they are left to fester or are kicked into the future for someone else to deal with. Climate change is the most obvious one. Many of the opportunities we reach for founder for lack of funding or of the political and social will to carry them out. They are among the grandest, boldest, and most brilliant ideas humans have ever had, confronting some of our biggest challenges. With a trillion dollars you could make them happen. At the very least, you could have a lot of fun trying. Such was my viewpoint as I set out to write.

And then coronavirus hit.

While it is sadly still a daydream that I personally have a trillion bucks to spend, that sum is no longer fantastical; it is valid, even vital, to investigate how it might best be spent. In March 2020, the United States Congress approved an economic stimulus package worth $2.2 trillion, aimed at mitigating the impact of the coronavirus. In November 2021, President Joe Biden signed into law a $1.2 trillion infrastructure bill, with part of that money earmarked for investment on electric vehicle charging stations. Biden also supported the $2.2 trillion Build Back Better bill, which would address climate change through massive investment in clean energy, among other things—things that I address in this book. The figure of a trillion dollars became almost commonplace, so much so that the idea of minting a trillion-dollar platinum coin at the US Treasury was revived.

Away from the US, in 2020 the leaders of the G20 group of nations agreed on a $5 trillion fiscal policy stimulus. The European Union passed a €1 trillion economic rescue package. In June 2020, the International Energy Agency estimated that governments worldwide would be spending $9 trillion in a matter of months on firing up their post-pandemic economies; another estimate put that figure at $12 trillion.[4] In 2020, globally, more than $6 trillion was created through quantitative easing.[5]

The money is there; huge sums have always been "there," in the global economy. But right now, projects on a trillion-dollar scale are being rolled out. Tens of trillions of dollars in economic stimulus packages are being chopped up, partitioned, allocated, siphoned. What if we could spend that cash? If only we could divert some of it, scrape a bit here and there from governments and banks, or quantitatively ease a trillion dollars into existence and spend it before anyone noticed. Imagine the possibilities. Imagine what we could achieve. What, say, could the World Health Organization (which has an annual budget of just $4.8 billion) do with $1 trillion for a global SARS-CoV-2 vaccination and treatment campaign? Or if the Intergovernmental Panel on Climate Change (IPCC—annual budget $200 million) was given this sort of money and told to spend it on mitigating the impact of global warming? A trillion dollars could really move the needle. That's what this book is about. We've seen in the response to the coronavirus that money can be found. And we've seen over the months of lockdown and social distancing that civilizational changes can be made. Indeed, we're beginning to recognize that they must be made. Writing this I often thought of Lin-Manuel Miranda singing, in *Hamilton*, "I'm not throwing away my shot." This, the shock and chance for reset that coronavirus has given us, is our shot. The victory of Joe Biden in 2020 makes dramatic change more possible.

But first let's set some ground rules. You know the movie *Brewster's Millions*? Richard Pryor's character has to spend $30 million in 30 days in order to inherit a $300 million fortune, but isn't allowed to own any assets at the end or give away the money. Under the rules of Project Trillion, spending must be broadly directed to saving both humanity and the planet. It can go toward improving human welfare, protecting and restoring the environment, advancing science and increasing our stock of knowledge, but it may *not* be used to form a new nation state, nor destabilize existing ones, nor indeed be spent for military, media, political or investment purposes, including fiscal stimulus. It would be tempting to buy Fox News (value: $20 billion) and repurpose it as a politically independent media operation, or to spend billions lobbying for (say) investment in renewable energy, and supporting politicians willing to stand up to fossil fuel companies. For that matter we might want to use the trillion as a fiscal weight on the scales to force the introduction of a carbon tax. I also toyed with the idea of forming a religion. Another time, perhaps. But I wanted to try and keep Project Trillion manageable. We'll have enough to think about by limiting the spending to merely protecting the future of humanity and all life on Earth.

We're sitting on top of a pile of money. Floating on an ocean of cash. In each of the following ten chapters I pick a mega-project— or often a collection of projects—and see how a trillion dollars could make them real. This is a personal list, a mixture of solutions to the world's biggest and most pressing problems, and things I'm excited, moved, and exercised by. There are projects that the world's top scientists are working on and problems that, for the sake of the world, desperately need solving.

The clock is ticking. Let's get spending.

An Ethiopian girl learning Amharic at a remote school, with no electricity inside the classroom.

1

Level Up Humans

AIM: To eradicate world poverty. Specifically, to lift the global millions living in extreme poverty above the poverty line, to break them free of the poverty trap and to set them up for a lifetime at a level above $2 per day subsistence.

DURING MY RESEARCH FOR THIS BOOK, only one of the dozens of people whom I spoke to—as it happens, a Harvard professor—refused to play along with the premise: to spend the money on *things*. "You should give the money to the poor," he said. "Yes, but it's a thought experiment," I countered. "What if we couldn't give it away and *had* to spend the money on, say, a science project." "No, it's morally wrong and you should give it to the poor," he insisted. It was as if I really had the money and was about to spend it on something he disagreed with. At first I was frustrated, but then I thought, okay, let's see. What *would* happen if we gave it all away?

When people are asked if we should give away public money, the reply is often: "But won't people just waste it?" Sure, they might. But questioning how poor people will spend money is effectively the same as asking if we should try to raise their incomes in the first place. And the answer to that question is undoubtedly yes. As we emerge from the crisis of coronavirus, the way we rebuild the world has to be green and sustainable, but it also has to be inclusive and leveling.

Even with the huge economic challenges of the global pandemic, we're living in the richest society the world has ever known. Our resources, and here I mean ours as a society not ours with the trillion bucks in our back pocket, are far greater than those of the richest of emperors, queens, and chieftains of the past, and far beyond the imagination of most of the billions of people who have ever lived. If we have the means, we should try to raise people out of poverty. It's as simple as that.

Or so it seems simple until you try to do it. Do we build roads? Sewage systems? Should we subsidize education? Pay for better reproductive health care for women? Improve nutrition? Should we just buy all poor people a cow and be done with it?

In fact, some aid programs do purchase cattle for people. Sometimes it doesn't have the best outcome. Not everyone wants a cow. It's a pain to feed and water and house a cow. People say, "Listen, thanks and all but can I just have the money that it cost to *buy* the cow, and *I'll* decide how to spend it?" Cows aren't what economists call fungible assets. They can't easily be exchanged. And they aren't climate friendly, either.

It's often the same with food aid or medical supplies. Bags of flour and sugar are very welcome in the event of severe famine, but otherwise people would prefer seeds—preferably seeds of crops bred to grow in the local conditions. "Even better," they say, "just give us the cash." People given mosquito nets in well-meaning malaria-control programs may end up using them for fishing. Emergency packets of Plumpy'nut peanut paste (the nutrition-packed supplement to fight severe malnutrition), malaria nets for the bed, and even pumps for the village well, may all be very useful under the right conditions, but people would still rather have the cash.

～

THE IDEA TO GIVE MONEY AWAY, not so much as a form of charity but as an improved way to run a society, has a long history. It starts with the idea, first floated by Thomas Paine in 1797, that landowners should pay an inheritance tax that is used to fund a basic income for everyone. Over the years many other thinkers, writers, and politicians have played with the idea that all citizens be paid a set amount each month, regardless of whether or not they worked. Amazingly enough, a similar proposal, the negative income tax, was almost signed into law under the Nixon administration in 1969 but was voted down by Democrats, who deemed the payments too small.

Since then, wealth inequality in the United States has grown to staggering levels; economist Thomas Piketty said it "is probably higher than in any other society at any time in the past, anywhere in the world."* The US is the richest nation the world has ever seen, yet it has higher poverty levels than any other Western democracy. What if Nixon had got his negative income tax bill through the House? (The game of "what if?" is a fruitless one—I often go back to the "what if" of the Florida recount in 2000 and the election of George W. Bush—but the Nixon "what if" is similarly intriguing and dismaying.)

Even before coronavirus hit, the idea of UBI—universal basic income—was being floated by a range of backers as diverse as Charles Murray of the right-wing American Enterprise Institute,

---

* There is some debate about this claim. See "Economists are rethinking the numbers on inequality," *The Economist*, November 28, 2019.

Mark Zuckerberg, Elon Musk, Hillary Rodham Clinton, and Black Lives Matter. When the pandemic changed the world, the calls for UBI were renewed. A guaranteed income would, say supporters, cushion the economic impact of the virus, and even slow its spread, because many workers would not be obliged to return to work when ill. Some UBI-style payments were made in Ireland to people made unemployed by the pandemic,[1] and in the US, three stimulus checks totaling up to $3,200 were made to millions of citizens. House representative Alexandria Ocasio-Cortez saw the crisis as an opportunity and called for another look at UBI.[2] Support for UBI was buoyed by the results of a trial in Finland in 2017 and 2018, where 2,000 people received unconditional monthly payments of €560. The results found that people with the payments worked six days *more* over the two-year period than a control group of 173,000 people on standard unemployment benefit, and that the UBI recipients also scored higher on financial well-being and mental health.

It's expensive, however. Entrepreneur Andrew Yang, a one-time candidate for the 2020 US presidential elections, proposed a $1,000 monthly payment to all adult Americans. That's a nice idea, but it would cost $2.8 trillion a year, and the federal government's total annual spend is only $4 trillion, so it's tricky to see how it would work. And it's clearly not for us. We are richly endowed but even we don't have enough money to start a universal income scheme for the US, let alone the entire world. So, if we want to give our money away, we need to think of a different rationale.

Let's do some back-of-the-envelope sums. If we divide the $1 trillion equally among the world population of 7.7 billion, each person would receive the (largely) non-life-changing amount of $130. One of the big objections to universal basic income is that people vary in the amount they have to begin with. If we did start a $1,000 scheme, we'd be giving that cash to people in poverty, but also to billionaires. So, for simplicity and efficacy, let's exclude people from developed countries from our arithmetic. My justification, by no means watertight, is that people in poverty in the United States and Western Europe will mostly not die of malnutrition and disease. I don't want to underplay poverty in Western countries. Poverty in the US, for example, is measured by household pretax income, and if a family of four brings in $24,339 or less they are classed as living in poverty. That's about 40.6 million people. The US has malnutrition and disease that is dragging down life expectancy, but there isn't starvation like in parts of Africa and South Asia and there are 607 billionaires in the country.[3]

Of course, there are billionaires everywhere these days. The richest person in Nigeria, Aliko Dangote, is worth $10.4 billion. But such is the extent of poverty in Nigeria that even if Dangote were to decide one morning, perhaps after being visited by the Ghost of Christmas Future, to give away all his money to the poor of his country, it wouldn't make much of an impact. There are 90 million people in Nigeria in extreme poverty, and they would each get $115 of Dangote's wealth. India's richest person is Mukesh Ambani, worth around $56 billion (the money comes from oil),[4] and again, even if Ambani was struck by a sudden Zuckerbergish urge to give away 99 percent of his wealth, it would not solve the poverty problem in his country. Later in the book we'll come back to what the billionaires could do if they all became infected with philanthropy, but for the purposes of this chapter, it seems clear that the problem of alleviating poverty in the developing world is greater than the poverty that exists in the West. We have to draw a line somewhere.

If we do exclude people in developed countries and divide the $1 trillion equally, that still only gives each person $161. So then let's exclude people earning above a certain amount. According to the World Bank, about 10 percent of the world's population, or 760 million people, earn $2 or less per day. (This proportion, by the way, the proportion of people in extreme poverty, is the lowest it has ever been in human history.)

If we divide our $1 trillion equally among these 760 million people, each one would receive $1,315. It's a tidy sum by any reckoning, and a life-changing amount if you live in extreme poverty. Could we really do that? I fretted about the responsibility of having this money; wouldn't it be irresponsible and wasteful to chuck it away like this?

~

SOME CHARITY PROJECTS and state-funded research have, it turns out, looked into just this scenario, and a growing body of evidence suggests that cash transfers are the most effective and efficient means of lifting people out of poverty. Charities such as Give Directly, as well as state governments running welfare schemes, have tested a range of different methods.

Sometimes people are just given money in a lump sum; or they may receive a smaller payment each month for a year, or over a longer term. Sometimes the money is given unconditionally, other times the payments come with instructions—it must be spent on children's education, for example. Sometimes everyone in a village gets the money; other experiments have looked at what happens when only the women receive it.

Let's have a look at what is known. First, about the classic question: Won't people waste the money?

If poor people can rely on an unconditional income, say the doubters, they'll just gamble and fritter it away. They'll spend it on alcohol or tobacco or other naughty things that economists call "temptation goods." Such is the common expectation. In Kenya, contributors to a charity expressed concern that the money would be wasted on alcohol consumption. In

Nicaragua, a government official suggested that "husbands were waiting for wives to return in order to take the money and spend it on alcohol."[5] One is reminded of William James's psychologist's fallacy: the projection of one's own concerns onto those of others.

Fallacy or not, the concern is widespread, which is one reason aid agencies support impoverished and environmentally impacted communities with goods and services rather than cash. Rather than sit in Washington, DC, and wring their hands over the matter, the World Bank decided to examine the actual consequences of cash transfers—to see what really happens when people are given money. The Bank conducted a thorough review of thirty studies of cash transfers to poor households in Latin America, Asia, and Africa.

The review found that almost universally the money was not "wasted" on frivolous or indulgent temptations, and often people spent *less* on temptations when they received the extra money.[6] The authors of the review, David Evans and Anna Popova, concluded that the evidence is strong that cash transfers are not wasted on cigarettes or alcohol. "We do have estimates from Peru," they admitted, "that beneficiaries are more likely to purchase a roasted chicken at a restaurant or some chocolates soon after receiving their transfer." But Evans and Popova hoped that even the most puritanical policy makers would not begrudge the poor a piece of chocolate.

⌖

WELL, OKAY, SO THE MONEY ISN'T WASTED. But if we are giving away all our cash, we want to ensure that it makes a difference. That it changes people's lives permanently, that it doesn't just lift them for a year, but rather sets them on a new course. There are studies into that, too.

I was frivolous about cows before—about programs that have given away cows. That's partly due to initiatives such as India's Integrated Rural Development Program (IRDP), which was criticized for ineffective and poorly targeting cow-giving. But here's a program that worked. A large randomized trial in Bangladesh engaged more than 21,000 households in 1,300 of the country's poorest villages, over seven years. In many such villages, the poorest women have only one option by which they can earn money, and that is through seasonal, temporary labor in the fields, or through maid services. These jobs pay, respectively, around $0.34 and $0.27 per hour. A rung up the poverty ladder, slightly wealthier women rear livestock, which pays $0.72 an hour. Since seasonal labor is temporary, women stuck with this option work two months less each year, reinforcing their poverty. It's one example of the poverty trap that researchers and aid workers see all over the world.

The Bangladesh trial identified ultra-poor households and gave them both assets and skills. That is, a small amount of cash, assets in the form of livestock (usually goats or cows), and the knowledge and ability to raise said livestock. The welfare donation came to the value of $1,120 per household over two years. Follow-ups after four years and seven years showed that women in the households that had been randomly selected for the intervention were more likely to be engaging in jobs that paid higher returns. Women spent 217 percent more time in livestock rearing, and increased the days they were working by 22 percent, showing that there was spare work capacity that could be taken up. Compared to the women in the control villages, women in the asset-transfer program were 14 percent less likely to come in below the $1.25 extreme poverty line. The one-off program had sustainably changed people's lives, and put them on a new trajectory out of extreme poverty.[7]

After the positive outcome of the Bangladesh trial, a similar program was tested in Ethiopia, Ghana, Honduras, India, Pakistan, and Peru, involving 10,495 people.

Again, the idea was that ultra-poor households are given an asset. They take advice and can choose sheep, goats, chicken, or cattle, for example. The options varied according to custom and country. In Ethiopia, people most often chose sheep or goats; in Peru, guinea pigs, which are farmed and sold as food. The cash-flow consequences also differed, because some people may immediately sell their asset, while others might keep it for longer-term revenue. The households were also given cash stipends and support if they required training. Despite these variations, and the big cultural differences in the different locations, all the trials showed profound improvements in the lives of the ultra poor compared to control groups. After the second year of the program, families enrolled in the treatment groups had greater assets, greater food consumption, better physical and mental health, and higher political engagement and women's empowerment. The basic idea of the program, that it provides a "big push" to escape the poverty trap, seems to work.[8]

In another study, in Sri Lanka, men were given a onetime cash donation. Five years later, the men's income had increased by 64 percent to 96 percent of the grant amount.[9] In Uganda, a study involving poverty-stricken, war-affected women given $150 and five days of business training showed good returns on the investment.[10]

❦

THE GIVE DIRECTLY CHARITY has been funding and piloting many UBI and cash transfer experiments, and following up their programs months and years after the initial donation. Here are some of the responses from recipients in a rural village in Kenya: "My son now is less stressed about me going hungry because I will get money to buy food in case he fails to send me money"

(Dorcus, 87); "I have been able to add capital to my business and increase the stock. This has increased my sales and improved my profits" (Irene, 23); "Since I went for check-up after receiving the transfer, my health situation has improved and I am able to go about my business without much stress" (Grace, 68).

These direct-giving programs are not perfect. Setting up a business in a remote village is not easy and competition for customers with fellow villagers can lead to tension. Some people report that the money is not enough and that there are conflicts within households over who is in charge of the money. But such hiccups are to be expected.

In another village in the Rift Valley, recipients of Give Directly cash transfers have used funds to install solar panels in their home. Another popular option is to use the money to fit a tin sheet to the roof of the mud-hut home rather than a straw roof. A metal roof lasts at least ten years, is safer and cleaner, and can be used to collect rainwater, which saves people the long trek to fetch water for irrigation. A straw roof must be replaced once or twice a year at a cost of $100–150, including materials and labor, which for someone on just 65 cents a day is prohibitive.

∾

THERE ARE MANY THINGS we don't fully understand about giving money away. Putting aside the microeconomic effects—that is, how individuals spend their money and change their behavior after cash transfer—we need to know what the impact is on the wider economy. There has been some research looking at the macroeconomic effects of giving out lump sums on inflation and job creation and public finance; they seem to not knock the economy out of kilter. But the sort of impact we might have, showering money on entire countries, is very different from the relatively small scale (to the tune of millions of dollars) of experiments conducted so far.

We need to pay a lot of attention to these impacts, Johannes Haushofer of Princeton University told me, to the effect on what economists call general equilibrium, the balance of supply, demand, and prices. There are some real-world examples we can look to. When Haushofer's team gave $1,000 in onetime cash transfers to more than 10,500 poor households across rural Kenya, there was a predictably large impact on consumption and asset acquirement, but little in the way of inflation.[11]

A rather different example comes from Alaska, where since 1982 all residents have received a yearly dividend derived from oil revenue. In 2015, this amounted to $2,072 per person. Damon Jones of the University of Chicago compared the effect of this cash transfer on control states—"synthetic Alaskas" made by drawing from other states in the union with comparable populations and assets, such as Wyoming and Washington. He concluded that there was no negative effect on the rate of employment in Alaska of the cash dividend.[12]

In Norway, the national oil fund (currently worth $1 trillion) is invested to provide a state pension but is also used to smooth yearly fluctuations in the government budget. We might tentatively hope not to fuel inflation or unemployment, but this will need careful monitoring: In rural areas, giving cash may contribute to inflation.[13]

Much has been made of the extraordinary reduction in global poverty over the last couple of decades. But, as Ola Rosling of the Gapminder Foundation put it, the last people—that last 10 percent—those are the hardest to reach. They are the poorest, living in the remotest regions. In papers describing the economic situation of developing countries, "remote" and "rural" are code words for "desperately poor." Most of the villages in that Bangladesh trial, for example, were in the *monga*, the famine areas of the north, the poorest parts of a poor country. Only 7 percent of the women engaged in the trial could read and write.

Anti-poverty programs have had great success, and the World Bank and United Nations say they can reduce extreme poverty to less than 1 percent by 2030, but it will take new initiatives, such as cash transfer, to reach the stragglers.[14]

⌒∾

HERE'S ANOTHER IDEA. People living in abject poverty can see no way out. They literally can see no way out. Their ancestors have often lived in the same way, and no one they know has ever escaped. If you are poor you can't plan, you can't hope, you can't dream of anything better. You just don't know any different. A lack of role models doesn't help. But what if you give them not cash, but something to aim for? This is a new kind of program called Aspiration Intervention. Kate Orkin at the University of Oxford told me about a project she has trialed in Ethiopia.

In a remote (yes, there you go) village east of Addis Ababa, in 2010 and 2011, thousands of households were invited to take part in a simple project that consisted simply of watching a video. It was an hour-long documentary about people in a similar village, showing how they had succeeded in setting up a small business selling food, or started a small agricultural operation. The video, which was really actors playing parts written for them, recounted how the people had moved up a rung or two on the poverty ladder. They had gone from poverty to relative affluence. The actors explained what they'd done, the rudiments of their business, and the steps they had taken to achieve their goals. In a control group, villagers watched an Ethiopian TV show and, in another group, villagers were merely interviewed.

Six months later Orkin and her team went back and interviewed the villagers. Those who had watched the aspirational video had 20 percent more children in primary school and spent 28 percent more of their income on education, compared to the control groups. They had more money saved and reported working harder on their land. Five years after that, the results persisted, with control families spending nearly 40 percent more of their income on the children's education, working more per day and investing more in products related to farming, such as seeds and fertilizer. All from just watching a one-hour film.

That's not to say that videos or other kinds of psychological intervention should replace material intervention; the effects are small in magnitude compared to those from economic interventions. "But it highlights that living in poverty affects people's beliefs about what is possible for them," says Orkin. "Intervening to encourage people to believe in what they can achieve and to have the self-confidence to try may have long run effects, and may form a key part of our anti-poverty toolkit." MIT economist Abhijit Banerjee, who shared the 2019 Nobel Prize in economics, has made similar arguments.[15]

Banerjee, incidentally, is one of a group of economists who looked at the effect of cash transfers in Kenya during the COVID-19 pandemic. People given money (75 cents a day) were less likely to suffer hunger or report sickness, including depression. The effects are modest, but it supports the idea that UBI can improve people's resilience in tough times.[16]

＄

THE PROGRESA SCHEME in Mexico in the late 1990s was at the vanguard of cash transfer initiatives.*[17] Mothers in poor households were given cash sums on condition that their kids attended

---

* The scheme, now known as Prospera, has had its critics and has now been scrapped by the Mexican government after twenty-one years.

school regularly. The scheme worked well, improving the educational outcomes for thousands of children, and has inspired many other experiments, such as a countrywide program in Brazil called the Programa Bolsa Familia (PBF). The PBF has helped to reduce financial inequality in the country by 15 percent, and extreme poverty has shrunk from 9.7 to 4.3 percent of the population. As a bonus, cases of infant mortality caused by malnourishment have also halved.

PBF payments are not universal. They are made only to families earning under a certain amount, but in Brazil in 2015 that was still a quarter of the population—almost 52 million people. In Peru, there was a cash transfer scheme that came with conditions. In enrolled villages, every two months the female head of the household received $143 if she had been sending the kids to school, taking children under five for health checks, and obtaining identity cards for the children.

One criticism of conditional schemes is that they are expensive to administer. Means-testing the families and checking up on the mothers accounted for 60 percent of the cost of the Mexico scheme, 50 percent of the cost of a similar one in Nicaragua, and 31 percent of one in Honduras. An alternative is to give money and just *label* it as "for educational use." Effectively, the parents are then trusted to spend the money on their children's education and not on alcohol or cigarettes.

In Morocco, the ministry of education tested this approach, giving relatively small cash transfers to families in poor communities, either "labeled" to be spent on education, or tied to a conditional cash transfer scheme that was subject to checkups. The team assessed the outcomes for more than 4,000 households in the scheme, including some 44,000 children, and found a 70 percent decrease in the rate of school dropouts. It made almost no difference on educational outcome whether the families were signed up to the labeled or the conditional transfer, but the saving on admin was profound.[18] Saving on administration, by the way, is similar on a smaller scale to why some thinkers on the political right support universal basic income: because it does away with layers of government welfare and social security programs. Political scientist Charles Murray, for example, suggests that if there was a UBI system in the US, then the government could scrap Medicare and Medicaid and require recipients to spend some of their UBI on health care insurance.[19]

"You might want to make your cash transfers conditional on something," advised Ben Moll of the London School of Economics. "Otherwise, you might just cause a blip and they go back to how they were before." The key is to avoid the blip, and to do that we need to identify the poverty traps, the pitfalls that keep people under the poverty line.

THERE'S ANOTHER WAY it is critically important to label at least some of our cash with conditions. This chapter is about leveling up humans, and has mostly been concerned with poverty in a purely financial sense. But there is another inequality, that of education. A focus on education might be the most lasting way to solve the problem of poverty. For example, in Indonesia between 1973 and 1979 the government built 61,000 new schools, doubling the number in the country. Nearly 40 years later, a study found the project had led to huge, positive intergenerational effects.[20]

We saw earlier how women in the Bangladesh trial were mostly illiterate. It's a similar story in many other poor countries: only 8 percent of girls in sub-Saharan Africa finish secondary school; in South Asia less than 50 percent go to secondary school. Globally, it is more likely that girls miss out on education than boys, and currently around 130 million girls are being excluded from school.

The Brookings Institute found that a woman who has never been to school has four to five more children than a woman with twelve years of education.[21] Girls who have been to school earn more, are less likely to marry as children, are less likely to have HIV or malaria, and farm more productive plots of land, which results in better-nourished families. In chapter 7 we will look at the price of carbon, and at how much it costs to get rid of a ton of carbon dioxide from the atmosphere. We usually think in terms of planting trees or even of technological methods to suck greenhouse gases from the atmosphere, but educating girls could be the most cost-effective of all methods, the single most powerful thing we could do to lift people out of poverty and tackle the threat of climate change.

The United Nations estimates that just an extra $39 billion per year could ensure universal education in low- and low-to-middle income countries. (The UN currently spends $13 billion on international aid projects for education.) Universal education, for just $39 billion. It's a shockingly small amount to not only ensure something that is a basic human right, but that can have an outsized effect on climate change. Project Drawdown, an organization providing resources to get the world to a carbon-neutral state, estimates that education of girls and the subsequent reduction in family size could lead to emissions reductions equivalent to almost 66 billion tons of greenhouse gases by 2050.[22]

We should immediately allocate some of our money to education. Such is the importance of this aspect of leveling up that it is tempting to assign money to ensure ten years' worth of universal education. That's around $400 billion.

The rest of our pot goes into cash transfer schemes. Giving money away feels good, right? Even when it's a trillion dollars we don't have.

Achieved

A ten-year period during which everyone in the world is lifted out of poverty and everyone in the world completes their education; the possibility of a sustained period in which poverty recedes, human ingenuity and potential is unlocked; welfare globally goes up a level.

Money spent

Universal education:	$400 billion
Cash transfer schemes:	$600 billion
Total:	**$1 trillion**

In vitro production of
antimalaria drugs at a
laboratory in Bukavu,
Democratic Republic
of the Congo.

2

Cure All Disease

AIM: To protect humanity from the next pandemic, create a new field of human biology, transform the human experience by curing, preventing, or treating all known diseases (not entirely excluding death). To avoid, if possible, splitting the human race into two species.

THE YEAR 2020 TURNED OUT to be pivotal for humanity—but not in the way many of us had hoped. There was an expectation that it was going to be the year we got serious about climate change, pulled together and rode a wave of changes that would build to a revolution in economics, agriculture, and energy; a civilizational upheaval that would lead to a better world. Instead, we got the pandemic that many scientists had been fearing for years.

The full human and economic impact of coronavirus, SARS-CoV-2, is still playing out. At the time of writing more than five million people have died, while hundreds of millions have had their lives disrupted or economically ruined. The economic impact is, at this point, incalculable—unquestionably many trillions of dollars, and still rising. But the tragedy is far from the worst it could have been. Unchecked, coronavirus could have caused 40 million deaths in 2020 alone, according to a report from the World Health Organization (WHO).[1] The virus itself could have been more virulent, and more deadly, and the miserable fact is that just because we've had coronavirus doesn't mean we can't get another even worse pandemic. SARS-CoV-2 has changed the world, and its tragedy will be felt for years, but for our purposes, in this chapter, we'll treat it as a shot across the bows. We need to use it to raise awareness of the threat of pandemic diseases. It also gives us an inkling of the threat to the world from climate change. Our response to coronavirus shows we can adapt and change our lifestyles, and from the point of view of this book it shows that governments *can* find money to spend—and particularly on public health projects.

We knew the risk posed by pandemics. Spanish flu killed between 50 and 100 million people worldwide in 1918. The UK maintains a risk register, a catalogue and assessment of the emergencies that could befall the country, and top of the list, at the start of 2020, was an influenza pandemic. Large-scale exercises in 2007 (code-named Winter Willow) and 2016 (Exercise Cygnus) showed what might happen to the health service, the economy, and the population if a 1918-style disease took hold. We knew what was at stake, and now we have firsthand experience.

We will invest, in this chapter, in preparing for the emergence of the next pandemic. And we will also think more broadly. In 2018, a disease infected 228 million people and killed around 405,000, mostly children under five, and mostly in sub-Saharan Africa. That disease, malaria, has been with us forever.

Malaria comes up all the time in world literature. In *The Tempest*, Caliban curses Prospero with the illness notoriously associated with marshlands and warm weather: "All the infections that the sun sucks up / From bogs, fens, flats, on Prospero fall and make him / By inch-meal a disease!" And before Shakespeare, references turn up in Chaucer in the fourteenth century, and Dante in the thirteenth. We have a medically confirmed case of malaria from ancient Rome, and before that references in Homer and Aristotle. The earliest known reference is in the ancient Chinese book of medicine, the *Nei Ching*, from around 2700 BCE, which discusses malaria and its symptoms.

The disease has killed perhaps half of all humans who have ever lived. Malaria is the world's greatest scourge, but it is preventable, and curable. In fact, we've done well: Deaths from malaria have been halved in the last twenty years. But still it clings on and, if we're looking for projects to stretch us and create a genuine legacy, its defeat of humanity's most deadly enemy has a certain ring to it. So that's coronavirus and malaria on our list. What else might we be able to eradicate with our windfall?

Tuberculosis is a bacterial disease that kills almost two million people each year, overwhelmingly in poor and middle-income countries. The factor holding back eradication has not been a lack of understanding the biology of the disease, but a chronic lack of resources, and the growth of resistance of the pathogen to our bacterial treatments. We can change this. We can tackle other tropical diseases too, such as schistosomiasis, a debilitating parasitic disease which affects 200 million people a year. But let's think bigger still. Picture a world free of *all* disease. Thousands of scientists and doctors are striving to treat and cure the world's biggest killers: cancer, cardiovascular disease, and neurological disease. Let's boost their chances, and see if we could transform the entire human experience by removing all illness.

⌒

WHEN HER DAUGHTER was still a baby, in September 2016, Priscilla Chan declared that she wanted to ensure not only that the infant would thrive but that she and her entire generation would grow up free from disease. "We'll be investing in basic science research with the goal of curing disease," she declared. Not merely curing breast cancer or Alzheimer's or diabetes or strokes—curing *all* disease.

Chan, a medical doctor who is married to Facebook CEO Mark Zuckerberg, was launching the Chan Zuckerberg Initiative (CZI), and announcing the investment of $3 billion into research aimed at preventing, curing, or managing all diseases by the end of the twenty-first century. It seems that most of the couple's Facebook stock (currently worth around $60 billion) will eventually be plowed into this project. Chan and Zuckerberg predict their efforts will cause life expectancy to increase to 100.* "That doesn't mean no one will ever get sick," said Zuckerberg. "But they should be able to treat it and manage it."

So people are doing what we are thinking of. On a smaller scale than a trillion dollars, to be sure, but it helps provide a framework. It helps us explore whether and how this is a problem that can be solved with money in this way. We can attempt to emulate Chan and Zuckerberg and Bill Gates, too. The Bill & Melinda Gates Foundation aims to eliminate many infectious diseases and has already saved the lives of more than 100 million children. But first let's break down the problem.

The Chan-Zuckerberg project focuses on the four main disease categories: heart disease, neurological disease, cancer, and infectious disease. These account for, respectively, 18 million, 9 million, 9.6 million, and 8.5 million deaths per year.[2] These figures are approximate, but you get the idea: There are many people dying, many of them preventable deaths. Furthermore, neurological

---

* Not to be outdone, the cofounder of Oracle, Larry Ellison, has vowed to "defeat mortality." "Death never made any sense to me," he has said. And PayPal cofounder Peter Thiel says death is just a problem to be solved.

disorders are the leading cause of disability worldwide, causing the loss of 276 million "disability adjusted life years" each year. A measure of ill health, a disability adjusted life year (DALY) is a year lost to disease or disability.

The CZI aim is to develop and build technology that will secure advances in these four key categories. This includes AI software to help image and interpret brain scans, machine learning to understand cancer from large datasets, tech able to provide continuous monitoring of blood to signal the earliest stage of disease—like advanced Fitbits—and a map of all the cell types in the body.

THE HUMAN CELL ATLAS aims to create an encyclopedia of the 30 trillion individual cells in the human body. Medical textbooks will tell you these cells are made of around 200 different types. You can probably reel off a dozen or so right away: liver cells, blood cells, neurons, heart cells, kidney cells, different kinds of muscle cells, those in the retina—and so on. But we're way short of the real number of types of cell. The retina alone is composed of at least one hundred different kinds of neurons. There are hundreds of different types of white blood cells making up our immune system, each performing slightly different roles in recognizing, labeling, and attacking invading organisms, such as bacteria and viruses, and cleaning up afterward. We have no systematic map of the different kinds of cells that make up our body. If we don't know our cells, how can we know ourselves?

This is not just an existential question. If we want to grow new tissues in the lab to replace diseased or damaged organs, we need to know exactly what cells to culture. If we want to tweak the genes of cells using gene editing, same thing. This is why the Human Cell Atlas project is such a big deal. Without this detailed map of our bodies, twenty-first-century medicine won't be able to fulfil its potential.

To take one example, there's a type of kidney cancer that is relatively common in children, called Wilms tumor. The treatment for it, currently, is chemotherapy. It's a grim, side-effect-laden experience for an adult, let alone a child. But analysis of the cells in the cancer show similarities to cells that haven't matured, so instead of killing them off with toxic drugs, it would be better to coax them into becoming mature kidney cells.[3] Or another example: You might assume that we understand cystic fibrosis pretty well, given that it is one of the few diseases that are known to be caused by a mutation in a single gene. But, after properly cataloguing the cells in the lungs and windpipe, scientists found that the disease acts through a completely new type of cell, called a pulmonary ionocyte.[4]

There are *many* other diseases and conditions where treatment will change when we know what is going on more accurately. Perhaps all treatments will change, it's that big a deal. We have the chance to shape this new field. It almost feels too small to call it medicine for the twenty-first century—you'd be seriously underestimating its impact. This will start an entirely new era of medicine that will make chemotherapy, antibiotics, and organ transplants look medieval.

When we shape this era, we will make it available to everyone; we'll earmark some investment for this. A quick reality check: you don't have to be living in a remote village in a low-income country to appreciate the scale of the task, or to be skeptical about claims to bring high-tech medicine "to everyone." I'm sure those of us living in high-income countries can all think of cases where doctors have brushed aside, misdiagnosed, or ignored symptoms of disease, with devastating consequences. It's going to be challenging to bring total health care to everyone in even one country, let alone the entire world.

In a vision statement for the World Economic Forum, Jenniffer Maroa, of the Department of Global Health at the University of Washington, says a world free from preventable forms of suffering can "easily" be achieved.[5] All it will take, she suggests, are new technologies such as Blockchain, the internet of things, and artificial intelligence (AI).

Is this enough? Can we tech our way out of this? As CZI sees it, in a very basic form the plan to cure all disease has three parts:

1. Bring together scientists and engineers.
2. Build tools and technology.
3. Grow the movement to fund science.

Point 3 is important. To cure disease globally you need to address at least two other major issues: poverty (which we've discussed) and climate change (which we'll focus on in the next chapter). For another, an investment of $3 billion over a decade is a nice chunk of change but is hardly going to cure all disease on its own.

Chan and Zuckerberg know this, and know that throwing money at the problem is not enough. The CZI is a limited liability company, not a charity. So Zuckerberg retains control of his shares and can invest politically, and in for-profit companies. Some critics have pointed out that by moving his stock into this new organization he can avoid paying tax on it and as a private company there is no obligation to be transparent about spending. That's a concern, but from our point of view, if you want to leverage your money to achieve an incredibly ambitious aim, it might be necessary. Because, whether it's a $600 million investment in Biohub, a nonprofit medical research organization in San Francisco set up by Chan and Zuckerberg, or $3 billion over a decade, or $60 billion over many years, it's a drop in the bucket when it comes to freeing humanity of all disease and extending everyone's life span.

And, if you want to make immense gains in public health on a global scale, and make them sustainable, there is one serious, ambitious, difficult, complex, and *expensive* thing that needs to be implemented. This doesn't seem to be something that is talked about or invested in by billionaires: universal health care.

∽

THE WORLD BANK PUBLISHED its first analysis of global health, the World Development Report, in 1993. Targeted at government finance ministers, the report showed that health expenditure could improve prosperity as well as individual well-being. To mark the twentieth anniversary of publication, an international *Lancet* commission put together an investment framework to achieve what they call a "grand convergence" in health by 2035. By this they mean bringing deaths from infectious disease in low- and middle-income countries, as well as child and maternal deaths, to the levels seen in the best-performing middle-income countries. These best-performing countries conveniently all start with a C, and are known as the 4Cs: China, Chile, Costa Rica, and Cuba. A grand convergence, the paper predicts, could prevent some ten million deaths in 2035.[6]

The team hammered out four key messages. First, the hard-cash economic argument that is most likely to be repeated in the corridors of power. The returns from investment in health care are big. By avoiding long periods of poor health, we increase the value of additional life years (VALYs, in health care acronym jargon), which generates an economic return that outweighs the health care investment we've put in by a factor of between 9 and 20. It's striking that—as we'll see throughout this book—spending what seems like a huge sum of money will often deliver big economic returns.

The second point is that the convergence is achievable in less than a generation. Presumably this means that investors can look to make money relatively quickly, too. Governments can be confident of balancing their books fairly soon after a large initial outlay. To be able to see an effect in a matter of years helps turn the convergence from a vision document to actionable policy.

The third point is that governments are underusing fiscal policies in health care. In other words, by increasing the tax on, especially, tobacco, but also on alcohol, deaths from noncommunicable diseases and from injuries can be sharply reduced in low- and mid-income countries. For example, a 50 percent price increase in cigarettes in China would prevent 20 million deaths and produce tax revenue of $20 billion annually in the next 50 years. The same price increase over the same time period in India would save 4 million lives and bring in an extra $2 billion a year in tax. As we'll see in the next chapter, reducing the subsidies paid to fossil fuel companies also has the effect of improving general health, mostly through the reduction in respiratory diseases.

But it's the fourth point that is most important for us: that universal health care is the most efficient way to achieve a convergence in global health. *The Lancet*'s framework was written before coronavirus, but the response of various countries to the crisis shows that universal health care is a good protector for pandemics, too.

Jeremy Farrar is director of the Wellcome Trust, one of the world's largest medical research charities, with an endowment of around $40 billion. As someone with real-world experience of problem-solving in global health, he is well placed to advise us on how to spend the trillion dollars. "The bedrock of your spending must be on universal health care," he told me. He was unimpressed by the size of *our* endowment: "It's not that much money," he said immediately. "To make real progress takes real money. If you compare it with other things in society, it's actually a very small amount."

Farrar emphasized that the bedrock of a push to improve global health has to be universal health care. "Nothing else is sustainable," he said. An equitable system of health care is necessary to improve maternal health, child health, to improve end of life care, and to fight epidemics. "Almost anything else is not equitable, not efficient, and will not deliver what you need sustainably."

By the middle of 2020, when there were more than 2.5 million coronavirus cases in the US, and when numbers were growing at an alarming rate, Cuba had reported just 2,448 cases in a population of 11.3 million.[7] One reason for their success in controlling the outbreak is likely to be the country's strong health care system. Cuba has 8.19 doctors per 1,000 patients, which is the highest ratio in the world.*

A trillion dollars isn't enough to change the world's health care system, so here's an idea. We allocate some of our money to instantiating a system of universal health care (UHC) in one country. This lucky country becomes a flagship, an advert to other countries of the benefits of UHC investment.

I want to choose a large country, so that the transformation is both a challenge and an impressive beacon. Vietnam, Indonesia, Nepal are all options, but I'm going with Ethiopia. It may also help in our planning that the current boss of the World Health Organization, Tedros Adhanom, is Ethiopian, and former health minister of the country.

With a population of 100 million, Ethiopia has a large economy but only about 3 doctors per 100,000 people. Maternal and child mortality is relatively high, mainly because most births take place in homes, without the presence of a trained modern

---

* Cuba's 2,448 cases work out at 21 per 100,000; by comparison, the UK stands at around 600 per 100,000.

midwife. Poor sanitation and inadequate nutrition exacerbate health problems. We should follow the example of Ghana, which Bill Gates says has the best health system in Africa. There are, for example, around 5 midwives per 1,000 births in Ghana, which helps bring maternal mortality rates down compared to Ethiopia. Ghana operates a universal service through its National Health Insurance Scheme. Ethiopia spends about $73 per capita on health care; Ghana spends $145.[8]

If we helped transform Ethiopia's health care system, of the many benefits, one would be that trained professional medics would be more inclined to stay in the country rather than emigrating. We'll need to look at and learn from elsewhere, including Indonesia's ambitious attempt to introduce a system of universal health care, Jaminan Kesehatan Nasional, which by 2019 covered 221 million people, or 83 percent of the population.[9]

So THERE'S SOME OF OUR MONEY to be spent on a demonstration of UHC. Another good chunk should go to vaccines, both their development and deployment. If this wasn't considered much of a priority before the coronavirus crisis, it is now. We all understand only too well now that the development, testing, and equitable distribution of a vaccine is a huge and costly undertaking, with results that are far from guaranteed.

The Bill & Melinda Gates Foundation works with Gavi, The Vaccine Alliance, to help fund vaccination in low- and mid-income countries, and this is something we can leverage. It is not easy, and not a case of just chucking money at the problem.

The battle against polio provides a good illustration of the task we face. The disease is caused by a virus that mostly affects children and which can lead to irreversible paralysis and sometimes death. The eradication effort has been a huge success, with cases dropping from 350,000 a year in 1988 to just 33 in 2018. The virus that causes polio was endemic in 125 countries and now it clings on in just two. However, the virus is highly infectious. A single infected child could lead to hundreds of thousands of new cases a year around the world. So it must be completely stamped out, just as smallpox was eradicated in 1980.

The wild polio virus is now endemic only in Afghanistan and Pakistan, mainly due to the instability and violence in those countries which prevents complete eradication, and because a pause in vaccination there during the COVID-19 lockdown allowed a resurgence. Nigeria, the only other country where polio was clinging on, was declared free of the virus in 2020.

We could incentivize the push to eradication by increasing the funding, safety, and resources for staff working in these incredibly challenging areas. In northern Nigeria, at least 67 health care staff working to vaccinate and eradicate polio were killed by gunmen suspected of being members of Boko Haram.[10] We will also need to improve the understanding of the safety of the polio vaccine to increase uptake. In Afghanistan, for example, we will need to counter the spread of anti-vaccine propaganda. In all countries, we will improve disease surveillance and data collection to better monitor the progress of eradication, and work to address people's basic needs—fresh water, food supply—in remote areas where the virus clings on.

The effort to eliminate the virus has prevented 1.5 million deaths and 18 million cases of paralysis. It's enough to want to ensure that no child suffers from the disease, but if you also want some dollar signs in your eyes, the estimated economic savings coming from eradicating polio are put between $40 billion and $50 billion.[11]

Polio disappeared from Ghana around 2003, and, also since then, no Ghanaian child has died of measles. For some diseases, the country has more than 90 percent immunity, thanks to the high level of vaccination. Vaccination uptake is so high because community clinics and health centers conduct outreach programs to get all babies vaccinated, telling parents about the benefits of the program.

This sort of positive, life-changing message is what we need to get out in countries where vaccination isn't universal. The WHO made a list in 2019 of things it considered major health threats, including the problem of what it diplomatically calls vaccine hesitancy.[12] The drop in the numbers of people willing to be vaccinated was a major concern even before coronavirus; it's that much greater now, in the Covid era.

Vaccine hesitancy produces a terrible toll. In Japan a fall in the numbers of people taking the HPV vaccine in 2013 (due to coverage of a cluster of adverse effects) is predicted to lead to 5,000 deaths from cervical cancer that could otherwise have been prevented. As well as exposing people to illness and death, vaccine reluctance and outright anti-vaccine propagandists prevent eradication of diseases.

⌁

THE WHO INCLUDED ON THEIR LIST, with admirable foresight, a threat labeled "disease X." Effectively they left a blank space that "represents the need to prepare for an unknown pathogen that could cause a serious epidemic"—a blank space duly filled the following year by SARS-CoV-2, which (almost certainly) crossed from bats and started spreading between humans.

Actually it was quite predictable that a serious disease would cross over from animals. Other examples include HIV, rabies, anthrax, Ebola, flu, MERS and SARS (both from the coronavirus family), and bubonic plague. All of these are zoonotic diseases that jumped from animals. A lesser-known zoonotic virus, Nipah, is of serious concern. It crossed from fruit bats to pigs to humans and was first picked up in the Malaysian village of Nipah in 1999. It has a shocking death rate of between 40 and 75 percent (compared to a 3 percent case rate for coronavirus). There is no treatment or vaccine, and if the virus mutated and became more easily transmitted between people—well, you can see the problem.

In chapter 4 we will look at how we can make changes to our relationship with the natural world. One consequence will be to try and reduce the chances of another virus crossing over. But our aim here is to deal with what we've already served up. We need an international institute for pandemic protection and response, perhaps working under the umbrella of the WHO.

In 2009, when the H1N1 swine flu pandemic threatened to take hold, the vaccines that were developed for it were snapped up by rich countries. Gavi subsidizes the cost of vaccines so poorer countries can afford them, and this is something we will do for any coronavirus vaccine, if it is not done by the goodwill of world governments (and we can't rely on that). We will also support the Coalition for Epidemic Preparedness Innovations (CEPI), an organization working on vaccines for many emerging diseases, including COVID-19. Thanks to CEPI funding, Robin Shattock, an immunologist at Imperial College London, was already working on the disease X response when coronavirus emerged. His team was able to design and start testing a vaccine in record time. Now, CEPI aims to reduce (or even end) the possibility of future pandemics by working to dramatically cut down on the time it takes to develop new vaccines. With $3.5 billion, CEPI believes it can trim future vaccine timelines to just one hundred days, as well as create a one-size-fits-all coronavirus vaccine, and build a library of prototype vaccines to hasten response time against future viruses.[13]

While $3.5 billion sounds like a lot of money, it will pay for itself many times over if we can avert the kind of disruption to our working life and economy caused by COVID-19.

⸺

WE CAN HELP BOOST VACCINATION RATES around the world, but we can also change the dial at a basic, research level. The development of vaccines is still largely based on a two-hundred-year-old technology, and we have vaccines for fewer than thirty diseases. As well as COVID-19, effective vaccines against HIV, malaria, and TB would be transformational. Then there are the more than 320 emerging infectious diseases that have been identified just since the 1940s.

We will fund new approaches to vaccine development, including RNA and DNA vaccines. These contain the genetic instructions to make proteins found on the surface of a disease-causing viral or bacterial cell. When the body is given the genetic vaccine, its cells read the instructions and make these proteins. This stimulates the immune system to mount a response, so provides protection if the real pathogen is ever encountered. Robin Shattock's putative vaccine is an RNA vaccine, as is a promising US COVID vaccine.

If we can create a universal flu vaccine, we would be protected from what is still one of the greatest health threats to our species: the emergence of a flu pandemic. It's the thing we were worried about when coronavirus came along, and perhaps the fact that COVID-19 was caused by a coronavirus and not an influenza virus is what wrong-footed the response of some governments. The development of a universal flu vaccine is therefore vital, and would help counter the rise of antimicrobial resistance. The evolution of superbugs immune to treatment, the problem of antibiotic resistance, is another item on the WHO threat list. Each year more than 1.5 million people die because the thing infecting them is resistant. It's a problem that could easily escalate.

We can also look to target noncommunicable diseases with vaccines. This has been successful with cervical cancer and liver cancer, but there are many other diseases that can be similarly targeted, including peptic ulcer disease, lymphoma, leukemia, and inflammatory bowel disease. Perhaps also type 1 diabetes.

We tend to focus on vaccination for children, but we can also look to invest in vaccination programs for the elderly. This will extend disease- and disability-free life. And here are some more dollar signs for you. "Every dollar invested in even the limited immunization we have in the world's poorest countries gives a return of $21, rising to $54 when broader societal benefits are included," says Seth Berkley, CEO of Gavi. "This is higher than any other health intervention." The impact on humanity would be nothing short of transformative. "A world without infectious disease would help to extend average lifespans, eradicate poverty, and boost economies worldwide," Berkley says.

We can add to this. Jessica Metcalf, an infectious diseases biologist at Princeton University, has proposed a Global Immunological Observatory, a global program of immune-system sampling from the general public that would allow scientists to pick up signs of new pathogens as they emerge.[14] SARS-CoV-2 will not be the last such threat, she believes. A Global Immunological Observatory will help "rapidly detect, define, and defeat future pandemics."

Gavi, which started work in 2000, has helped vaccinate more than 760 million people in low- and mid-income countries, and has saved more than 13 million lives. For the period 2021 to 2025, it required at least $7.4 billion to vaccinate another 300 million people—that was before factoring in the cost of coronavirus. We can sign off on this, and allocate $100 billion to establish a massive research and development program, an international virus detection, vaccine response, and development lab.

~

FOR MALARIA, we can add to the World Health Organization efforts. The WHO is putting $8.7 billion a year into its ten-year malaria strategy. If we added, say, $100 billion, what could we do? The classic idea in epidemiology is to eliminate the common cold by giving everyone a wonder drug at once. Bingo, cold gone. With our windfall we could try to ensure that everyone in need, throughout the world, received the best available antimalarial drugs (currently the artemisinin-based combination therapy, ACT). But, even with our money, we can't reach everyone in remote places in sub-Saharan Africa and, without also controlling the mosquito population density, malaria will come back instantly. So we'd need to fight the insect as well as the disease. Insecticide spraying and insecticide-treated bed nets are currently the frontline interventions, and have been driving the decline in malaria since 2000. But there is always this residual transmission from a few infected mosquitoes clinging on, and the disease comes back.

Malaria is caused by a parasitic microorganism, *Plasmodium*, carried by a mosquito. We've tried over the years to eradicate it, but, every time, the parasite or the insect—or both—have clung on, evolved resistance to our control methods, and bounced back. So we need to take out the mosquito reservoir, and to do that we need a method of beating evolution. Target Malaria, an international research consortium with backing from the Bill & Melinda Gates Foundation, is working on a solution. They want to use a method of genetic modification called gene drive, which causes female insects to become infertile, but for the first time it does so in a way that has proven immune to the evolution of resistance. Gene drive forces a gene to be transmitted to the next generation 100 percent of the time, not just with the usual 50:50 rate, which prevents natural selection from finding ways around it. The modified gene renders the females unable to produce eggs, and in lab tests populations of mosquito have grown and collapsed as the gene spreads. With a method like this, you need to introduce modified males, bred to carry the infertility gene, into the population. Rolled out on a big enough scale in the wild, with a huge modification and breeding program, we can eliminate local populations of malarial mosquitoes. A large trial using this gene drive method is currently underway in Florida to eradicate a different species of mosquito, the one that transmits dengue fever (for which there is no effective protection), following successful field tests in Brazil.[15]

If you are wondering about the ecological wisdom of eliminating an entire species from an area, it's a valid concern. But note that we have often tried to eliminate mosquitoes using chemicals, which have horrendous and damaging off-target effects, so removing one species without insecticide is an improvement on that. In addition, few animals rely on malaria mosquitoes for food: A study looking at the effect of getting rid of mosquitoes found that removal is unlikely to have a big impact on the local ecosystem.[16]

⁓

IF WE CAN END TROPICAL DISEASES, and generate a shift toward equitable, universal health care, and add to the boost in research and development being provided by programs such as the Chan Zuckerberg Initiative, and see breakthroughs in the treatment and prevention of everyday illnesses such as cancer, diabetes, and cardiovascular disease, then we might start to speed the global increase in average lifespan. It's already fast. The average person today can expect to live for 72.6 years, which is longer than people in the best country in the world (Norway) in 1950. But the average value conceals a big inequality, from an average of 53 years in the Central African Republic to 85 in Japan.[17]

In 2020 Mark Zuckerberg predicted that by the end of the decade, we would all be living longer lives. "Scientific research will have helped cure and prevent diseases to extend our average life expectancy by another 2.5 years," he posted on Facebook. That's not controversial. A 2017 paper in *The Lancet* assessing long-term trends around mortality and longevity predicted something similar, even without CZI investment.[18] If we could achieve the incredible feat of eradicating, preventing, or curing all disease, we would all be living to around 100, which is about the natural human maximum, give or take.

How much further might we go if we treated death itself as a disease?

Starting on March 27, 2015, and ending on March 1, 2016, Scott Kelly spent a year on the International Space Station—the longest time an American has spent in space. What was fascinating about Scott's trip was that his identical twin brother Mark, a former astronaut himself, remained on Earth. By comparing the physiology and genomes of the brothers, NASA scientists have been able to look at how the exposure to the space environment changed the activity and makeup of Scott's genes.[19] It appears that genes related to DNA repair were more active, for example—not surprising, given that the radiation exposure in space is greater than on Earth. More puzzlingly, the telomeres in Scott's cells appeared to get longer. Telomeres are threads at the end of chromosomes that tend to get shorter with age. We don't know what it means that they got longer in space, and when he'd been back a few months Scott's telomeres returned to normal length. Another concern is that in places Scott's DNA shows potentially destabilizing mutations that could prefigure cancer.

Chris Mason, a geneticist at Weill Cornell University in New York, studies the effects of spaceflight on human physiology, and is testing countermeasures and preventative treatments on mice. He says that in twenty years, when we've got more data, we'll be in a position to engineer humans for spaceflight and for survival on Mars. If we're off planet for any length of time, genetic changes are going to happen. "It's not if we evolve; it's when we evolve," Mason says.[20] So, the reasoning goes, we might as well gene-edit humans to help their physiology survive.

There is plenty of science fiction where humans break free of their Earthly shackles and expand across the galaxy—the *Culture* novels of Iain M. Banks, *Ancillary Justice* by Ann Leckie, and Kim Stanley Robinson's *2312*, to name just a few. In these books humans have also broken free of one of the key boundaries that define us: mortality. People live for thousands of years, getting repairs, upgrades, augmentations, and even new bodies when necessary. The idea that we might be able to change something as fundamental as the fact of death is a dizzying prospect. According to some, it's where we're going as a species.

An important step on this path was taken by He Jiankui, formerly at the Southern University of Science and Technology in Shenzhen, China. It was there that he performed one of the most controversial medical procedures of all time. He genetically altered two human embryos, implanted them, and allowed them to develop into babies—two girls—and be born. He was jailed and fined in 2019 for violating medical regulations. He says he was attempting to make the children immune to HIV by editing a gene called CCR5 to prevent infection. The technique that He used is called CRISPR/Cas9, and it allows genes to be located and changed or disabled. This kind of genetic engineering was possible before 2012, when CRISPR was invented, but it was very complicated, time-consuming, and expensive. Now, it is cheap and easy. It is far from clear whether He succeeded, or if the gene-edited girls will experience any health consequences. But the condemnation from the medical and scientific community was almost universal. We don't know if the procedure was safe; the parents were ill-informed and documents were forged. However, the genie is out of the bottle.

If we wanted to, we could easily set up a lab to do the same. If *I* wanted to in real life, even without the trillion bucks, I could do it. CRISPR is no longer so technically difficult that only professionals can attempt it: There are YouTube videos and internet guides available and it's on the verge of becoming a hobby activity. It's become cheap to assemble the equipment (a few thousand dollars) and I could even purchase human embryos. It's not illegal to buy the cells or the equipment, but it is illegal, at least in the UK, where I live, to edit the genome of a human embryo for reproductive purposes. The fact that it is so easy makes it hard to believe that people won't try. Even following the condemnation of He, Denis Rebrikov, a Russian scientist, has said he intends to gene-edit an embryo, implant it in a woman, and allow it to come to term and be born.[21]

I'm not tempted to sign off on a research lab looking at using gene editing to change the germ line, the sperm and eggs, so changes get passed on. We're not at the point where we can do this. Even if it was shown to be safe, the ethical impact of gene editing on humans is far from resolved. Look at the controversy when deaf parents ask to use genetics to ensure their children are also deaf. Legislation around IVF in the UK says that if an embryo is tested and found to have genes associated with deafness, it should be discarded. But to some in the deaf community that is discriminatory. Deafness, to them, is not an undesirable or debilitating trait, but a way of life and a culture they want to preserve. Some people might want to knock out genes associated with heart disease, or cancer, then others might want to adjust genes associated with light skin, or dark skin. Or short stature, or epilepsy, or autism, or homosexuality.

Let's say society came to allow gene editing for a narrow range of well-understood genetic conditions, such as cystic fibrosis, or heritable cancers or heart disease. That seems reasonable, but we might inadvertently start a speciation event because gene-edited people might not want to breed with "wild-type" humans, in case their offspring pick up a faulty gene all over again. You immediately create a two-tier structure in the human species. This is perhaps the sort of outcome you might get in the future with the large-scale editing envisioned for space travel.

What *is* currently being done is gene editing to cure diseases by making genetic changes that don't get passed on to the next generation. In 2020, two people with beta thalassemia and one with sickle cell disease had their bone marrow cells gene-edited to compensate for the mutated genes that cause their diseases. In beta thalassemia and sickle cell disease, there are mutations that affect hemoglobin, the protein that carries oxygen in red blood cells. Bad cases require regular blood transfusions, so what the researchers did was to activate a gene in the patients' bone marrow that started producing a kind of hemoglobin called fetal hemoglobin, which stops being made when we're born. The researchers took bone marrow cells from the patients, used gene editing to turn on the production of fetal hemoglobin, then put those bone marrow cells back in the patient. The cells start making fetal hemoglobin, and the patients no longer need blood transfusions. It was the first time that CRISPR gene editing was used to treat an inherited genetic condition in humans.

In this case, the changes were made only in the bone marrow, not in the eggs or sperm, so there aren't permanent inheritable changes, and they won't get passed on. But it seems inevitable that germ-line gene editing will be performed.

By investing in research into gene editing in an open and transparent way, we can at least try to make it safe. Our institute will follow WHO guidance and provide open access to all the tools and methods we are using, and we will enlist ethicists and minority groups to inform the research and decide upon acceptable targets for research.

We can also try to correct some of the inherent biases in genetic research. Often lost in the celebration of human genome sequencing is the fact that only 2 percent of the world's genomics data comes from Africans and people of African descent.[22] The huge European genetic bias means we fail to properly predict, diagnose, and treat diseases across the globe. This is something we can change.

All this said, I'd prefer to focus gene-editing investment and research in animal and food crops, as we'll see in chapter 8. At the moment, work to extend lifespan through gene editing is going on, but only in mice. Scientists investigate aging in lab mice that are bred with a genetic disorder that causes accelerated aging. Juan Carlos Izpisua Belmonte's lab at the Salk Institute in California used CRISPR/Cas9 editing to treat the mice. This method uses the Cas9 protein to very specifically knock out the bad genes causing the aging, without disrupting any other gene activity. After being given a virus carrying the Cas9 therapy, the mice became more active, grew stronger, and showed better cardiovascular health. Their lifespans increased by about a quarter.[23]

The idea is to eventually be able to treat the molecular causes of aging in humans. Again, we can support this sort of research, but our focus has to be on universal health care. Genetic treatments for aging are far off and for a long time will be available only to an elite group.

When I was an impatient youth, I felt an irritation with being born in what I perceived as a clunky analogue world. Computers were weak; you couldn't talk to them. We weren't a space-faring species. We were powering the world by burning fossil fuels. I'd been born too soon, I thought, to see the cool stuff happen. That feeling has thankfully faded with age. I appreciate now that an incredible amount of "cool stuff" has happened, and that the world we live in would have been almost unimaginable to my grandparents when they were having children in the 1950s. But I do want to see the future. What we do now will have long-term consequences.

It seems outrageous, hubristic, even, to suggest we can cure, prevent, and treat all disease by the end of the century. But Cori Bargmann, a geneticist and neuroscientist who heads up the Chan Zuckerberg Initiative, takes the long view. "Going back in time a similar distance, much of modern medicine would have been unthinkable," Bargmann says, "from organ transplants and deep brain stimulation to treating cancer by manipulating the immune system."[24] In eighty years, we will almost certainly have made changes that make our best medicine look like crude guess-work. We may even have reduced the incidence of illness and disease to that of extreme rarity.

But there's a caveat. If we don't rein in our carbon dioxide emissions and halt climate change, our bold plans for a medical and health revolution will be for nothing. The effects of global warming on mortality each year could become, by the end of the century, worse than those of all infectious diseases put together, according to a report by the Climate Impact Lab.[25] Getting ourselves off fossils fuels is the greatest challenge of our time, perhaps the greatest we've ever faced, and this is the challenge of our next chapter.

Achieved

Tens of millions of lives saved, average global lifespan increased by 10 years.

Money spent

Universal health care introduced in Ethiopia: . . . . $100 billion

Vaccine development; expanded immunization program for emerging diseases; vaccine outreach program: . . . . $100 billion

Eradication of malaria and other tropical diseases: . . . . . $100 billion

Eradication of tuberculosis: . . . . $23 billion

Cure and eradication of HIV and other infectious diseases: . . . . $30 billion

Antibiotic resistance: . . . . $10 billion

Complete map of all cell types in human body: . . . . $5 billion

Solving heart disease, neurological disease, and cancer: . . . . $100 billion each = $300 billion

Curing damaged cells in the body and extending healthy life span by 40 years: . . . . $200 billion

Total: . . . . **$868 billion**

Greta Thunberg holds a placard reading "School strike for the climate" outside the Swedish parliament on November 30, 2018.

Go Carbon Zero

AIM: To massively cut our emissions of carbon dioxide and wean the world off fossil fuel. To transition to renewable energy, as fast as possible, and rebuild the power grids across the world. To move to zero-carbon transport and industry, to massively boost energy efficiency, and to change our housing. This is a big one.

ON MARS, it drifts down through the thin atmosphere as snow, forming a frozen ice cap on the South Pole. On Venus, it makes up almost the entire composition of the atmosphere, pushing the surface temperature above 840°C, hotter than the melting point of lead. On Earth, it is a trace gas, making up a mere 0.4 percent of the atmosphere. But boy is it a player. If ever there was a molecule that punched above its weight, it's carbon dioxide (CO_2).

For more than 150 years, scientists have known that carbon dioxide in the air warms the planet. The American scientist (and women's rights campaigner) Eunice Foote was the first, in 1856, to show experimentally that "an atmosphere of that gas would give to our earth a high temperature."[1] As a woman in the nineteenth century, her discovery was mostly overlooked and an Irish physicist called John Tyndall tends to be credited with the first discovery of the heat-trapping effect of carbon dioxide, in 1860. Forty years later, Swedish chemist Svante Arrhenius suggested that burning fossil fuels might eventually increase global temperature. Politicians have also known for decades. In 1965, the US President's Council of Advisors on Science and Technology reported that greenhouse gas emissions would warm the planet and that this "could be deleterious from the point of view of human beings." In 1982, Margaret Thatcher warned the United Nations about the risks of climate change: It is "real enough," she said, "for us to make changes and sacrifices so we may not live at the expense of future generations." And so it continued. In 1988, NASA climate scientist James Hansen told the US Congress: "The greenhouse effect has been detected, and it is changing our climate now."

You know what happened. We didn't make sacrifices; we didn't slow down greenhouse gas emissions, made from burning oil, gas, and coal. We didn't even hold them steady, we accelerated them. In 1988, the world emitted 20 billion tons of carbon dioxide; today, the annual figure is 37 billion. Half of *all* the carbon dioxide we've emitted since the Industrial Revolution has happened in the past 30 years.

The coronavirus crisis has not changed things. Lockdown and restrictions on movement had a big impact on our lives and on our travel, but an analysis by Piers Forster and colleagues at the University of Leeds found this will have almost zero impact on eventual temperature rise.[2] Carbon dioxide emissions dipped, but not by enough, and not for long. But what *could* make a difference is investing in green economic recovery plans as we rebuild from the crisis. The world is in recession, but unprecedented money has been found. At the risk of sounding like a corporate video, this is an opportunity to create a cleaner world, to rebuild our lives so that we may have a better future.

In this chapter we will see what we can do to speed the transition from fossil fuels to renewable energy, to build a world we can live in. We have to ensure we get to a carbon-neutral or carbon-negative state, meaning a civilization that draws down as much or more carbon dioxide from the atmosphere as it puts in, by 2050 at the latest. Why? Because, not to put too fine a point on it, otherwise we're looking at a world of food shortages and droughts, extinctions and economic catastrophes, floods and forced migration. Or, rather (because these things are all *already* happening), a world where they are all far more common.

⌒

BY CONVENTION, carbon dioxide concentration in the Earth's atmosphere is measured from a point near the summit of Mauna Loa, the largest volcano on the planet and the heart of the island of Hawaii. Its remote location, altitude, and lack of vegetation make it the ideal place to get unbiased atmospheric measurements, and the Scripps Institution of Oceanography has been monitoring levels there since 1958. In 2020, the concentration of carbon dioxide reached 416 parts per million (ppm). We know from samples of air trapped in old ice that before the Industrial Revolution the concentration of CO_2 was around 280 ppm. That was the level the gas had stayed at for hundreds of thousands of years.

Its growing concentration is the primary driver of climate change, and the reason the average temperature of the planet has warmed almost 1.1°C (about 2°F) in comparison to preindustrial averages. There are other gases that are also driving global heating, including methane, nitrous oxide, and hydrofluorocarbons (HFCs), collectively known as greenhouse gases, but carbon dioxide is the most significant and is the one we'll concentrate on getting rid of.

The impact of rising seas, more frequent and widespread wildfires, stronger hurricanes, extremes of hot weather, of drought and flood have become regular items on news reports. It is just the beginning. The problems will get progressively worse as we add to the stock of greenhouse gases in the atmosphere. In fact, the next half a degree, taking us to 1.5°C (2.7°F) of warming, will have proportionately greater impacts. This "accelerating risk" means the half degree after that could have *exponentially* worse effects. An international team of scientists recently concluded that a rise of 2°C (3.6°F) or higher looks increasingly unmanageable, and a danger to natural and human systems.[3] The Intergovernmental Panel on Climate Change (IPCC) says that if we let global temperatures rise by 2°C (3.6°F), then *really* dangerous things start happening.

The last time the planet was 2°C (3.6°F) warmer, sea levels were 13 to 20 feet higher. Although we could see this 2°C (3.6°F) warming by mid-century, it will take centuries for all the ice to melt. But we'd be looking at the eventual submergence of London, Shanghai, Mumbai, New York, and thousands more miles of coastline. The destruction of most of the ports and docks and economic infrastructure that keeps the world running. We'd be looking at the total reshaping of civilization. In all likelihood we would see increases in disease and starvation, political instability, and war.

Set against this, the investment of a trillion dollars looks like an absurdly small amount. The economic impact of unfettered warming in the Arctic alone has been put at $67 trillion by the end of the century. If we allow warming of 3°C (5.4°F)—and this is where we're currently heading—then even the austere IPCC starts to use terms such as "catastrophic" and "apocalyptic." We would probably lose the Amazon rainforest, and certainly we can expect a global food crisis. This is the path we're on. Along the path there are tipping points in the Earth system, such that, for example, enough warming builds up to guarantee that all the ice in Greenland and Antarctica melts, locking in sea level rise for centuries to come even if we did manage to claw back some carbon dioxide later.

The impact of climate change is unequally spread both between and within countries, with the Global South hardest hit. Countries such as India will suffer despite having contributed little to the problem so far. In the United States, the south will suffer more than the north from heatwaves, drought, and sea level rise.[4] But we'll all feel it. Even countries sometimes thought of as out of the climate change firing line will feel it. In 2020, the wheat harvest in the UK was down by a third because of the year's heatwave.[5]

The bottom line for everyone is that we have to stop putting carbon dioxide into the atmosphere, and remove a lot of what we've already put in. Everyone (well, almost everyone) knows this. At the Paris climate talks in 2015, and again in Glasgow in 2021, world leaders agreed to try to keep warming to 2°C (3.6°F) and, if possible, 1.5°C (2.7°F). To do so would require huge cuts in carbon emissions, far bigger than the ones agreed upon at the talks, which the UN Environment Program said would heat the planet by 3 to 4 degrees this century. And, in the years since that "historic" accord, emissions have gone *up*. Greta Thunberg was right to turn down the Nordic Council's environmental award for 2019,[6] and she was right to denounce world leaders at the United Nations for their lack of action. "We will never forgive you," she said. "How dare you."

∽

THERE CAN BE A SENSE OF DOOM about this, especially if you look at reports on and by the fossil fuel industry. An analysis by the

Norwegian consultancy Rystad Energy, commissioned by the *Guardian* newspaper, forecast that the world's biggest oil companies are accelerating production, with a planned increase of more than 35 percent by 2030.[7] Much of this will come from oil operators in the United States, which makes me think we should focus a significant chunk of our investment there. But Global Energy Monitor reported that between January 2018 and June 2019 China had expanded the use of coal by 43 gigawatts, with another 150 gigawatts about to come online. That's about the same as *all* the coal used by the European Union—and on top of the country's already massive emissions of carbon dioxide.

So we'll look to focus our investment in the US and China, and parachute help in around the rest of the world. India, especially, needs to come off the fossil fuel path it is on. Over half of India's primary energy consumption comes from coal,* and if it continues to hitch its growth to fossil fuels its emissions will be so great that we will certainly miss the 2-degree target. Our trillion dollars alone can't do everything, but we have to hope it will have a significant domino effect, that the gravitational pull of our investment will attract more and more green investment and drag us to a world where we draw down more carbon than we emit.

There are positive signs that we are pushing with the tide of history. President Biden put climate action at the center of his campaign and even the big oil and gas companies understand they must move in the direction of sound business strategy and financial probity. In August 2020, the BP CEO pledged that the company will be net zero by 2050, and will cut oil production by 40 percent by 2030. (BP made this pledge while announcing a record loss in profits.) Then Xi Jinping announced at the UN that

---

* In 2017, 56 percent of India's primary energy consumption was from coal; in 2040, that figure is projected to be 48 percent; see *BP Energy Outlook: 2019 Edition*.

China would be carbon neutral before 2060. China is by far the world's biggest emitter of carbon dioxide, accounting for 28 percent of the global emissions, so this could really move the dial.*[8] South Korea and Japan both followed, pledging net-zero emissions by 2050.

Another reason for optimism is that there is big money to be made in transitioning to zero carbon. We've known that at least since economist Nicholas Stern's report in 2006, which showed it made strong financial sense to invest now in order to save later; a 2016 follow-up report only strengthened the findings.[9] A report from Stanford University concluded that limiting warming to 1.5°C (2.7°F) would save $20 trillion by the end of the century.[10] A 2018 IPCC report found that to keep us at 1.5°C (2.7°F) would cost approximately $2.5 trillion per year until 2050 in investment in the energy sector, and $775 billion per year in energy demand measures. It sounds like a lot of money—it *is* a lot—and much more than we've got. But the economic benefits of staying at this level of warming are four or five times the size of the investment.[11]

We have to speed the transition to renewable energy. That is, we have to massively ramp up sources of power that regularly renew themselves, such as wind, water, and solar energy (often referred to as WWS power). It's not "just" that we'll save ourselves money and reduce human suffering and environmental damage. The more we do now to reduce emissions, the more time we will buy ourselves before the *really* bad stuff happens.** There is a limit to how much greenhouse gas we can dump into the atmosphere before we guarantee warming the planet by 1.5°C (2.7°F),

---

* Analysts at Sanford C. Bernstein & Company suggested that to get to carbon zero by 2050 China would need to spend $5.5 trillion, or about $180 billion a year.

** Here's another example: Sea level rise threatens up to 630 million people by the end of the century; see Scott A. Kulp and Benjamin H. Strauss, "New elevation data triple estimates of global vulnerability to sea-level rise and coastal flooding," *Nature Communications* 10, no. 4844 (2019).

and a reasonably accepted figure for this is 628 billion tons of carbon dioxide equivalent (taking into account other greenhouse gases). That means, at the current rate of emissions (44 billion tons of carbon dioxide per year), we have about fifteen years.

Roll up your sleeves. If ever there was a mega-project to get behind, it is this.

∽

LET'S START WITH SOME GOOD NEWS. Harnessing and using renewable energy has become cheaper than ever, cheaper than many ever thought possible. Right now, in many cases it is cheaper to get electricity by building new wind or solar facilities than it is to *carry on* getting it from existing fossil fuel power stations.

Wind is in the lead here. It is so plentiful that, if we harness it, it can provide all the power we need. We've known this for years—a 1991 study showed that the wind of just Kansas, North Dakota, and Texas could potentially provide enough electricity for the whole of the United States.[12] Today, the US has a lot of turbine capacity already installed and Texas alone generates a quarter of the country's wind power.[13] Globally, wind was providing 597 gigawatts of electricity—around 6 percent of total demand— by the end of 2018.[14] But that's not nearly enough. We need to increase its share—massively and fast.

In the UK, in 2019, the government (which had previously blocked most onshore wind projects) signed off on wind farms that were projected to generate electricity more cheaply than existing gas-fired power stations in just four years. The new contracts show that it will soon be cheaper to build new wind farms than to keep existing fossil fuel facilities running. This drastic reduction in the price of wind power has come mainly through larger turbines, which capture wind from a larger area and provide a longer and more consistent delivery of electricity. In 2017, the world's biggest wind turbines started generating power off the coast of Liverpool. At 640 feet, the thirty-two turbines stand higher than the London Eye.[15]

As turbines get ever bigger, they will need to get lighter. One way to do this is to use superconducting rotors, which use fewer magnets than conventional designs, making them cheaper. The superconductors can carry current more efficiently, which also saves weight. Field tests of the first superconducting multimegawatt turbines have now been carried out.[16]

In Kenya, the largest private investment in the country's history led to the creation of the continent's largest wind farm in 2019.[17] This is encouraging, as countries in Africa need to be quickly converted to renewables because otherwise more coal-fired power stations will be built to supply the growing energy needs. A large investment is needed to get the ball rolling in poorer countries and we can provide that, both for wind and for solar photovoltaics (PV), where costs, as for wind, have been plummeting.

The costs of power from different electricity-generating sources can be compared by calculating what they cost to build and operate, their operational lifetime, and the amount of energy they supply. This gives the levelized cost of energy (LCOE).[18] There are various ways to calculate and estimate this value, but renewable energy, coming in at anywhere between 5 and 10 cents per kilowatt hour, is usually cheaper than fossil fuels or nuclear. Lazard, a finance-analysis company, produced a report on LCOE that showed that large-scale solar farms and onshore wind power are now consistently the cheapest forms of generating energy.[19] It has actually been cheaper to build solar and wind facilities than to build new fossil fuel or nuclear power stations for some years. But it is still startling to discover that it now costs less to *build* new renewable power plants than to carry on getting power from existing coal and nuclear.

The scale of solar has been rising fast, too. The world's largest solar power plant, the Noor Abu Dhabi facility, started sending power to the grid in July 2019, with a 1.2 gigawatt capacity. But it won't have the record for long, as the emirate is going ahead with a 2 gigawatt project in the Al Dhafra region. Meantime, a 2.6 gigawatt solar plant in the Mecca region of Saudi Arabia looks like it is getting closer to becoming a reality.[20]

⌒⌒

OUR MASSIVE INVESTMENT will be a huge shot in the arm for the renewables industry, but we have to ensure that economics will take over after we've given it a boost. And we also need to take a look at the world's electricity grids, and the ways and means of storing renewable energy.

The big concern with renewable energy has always been that you can't guarantee power when you need it. Even if it can be made cheap enough, the peaks of production—when it's sunniest and windiest—don't always tally with the peaks of demand, and it's difficult to store the surplus. This has been one of the major reasons for the reluctance of governments to invest in renewables, and it's been an explanation—an excuse, really—for some of the continued subsidies paid to fossil fuel operators: the idea that "renewables can't keep the lights on," whereas fossil fuel power stations can literally stockpile their fuel and burn it when needed. This is where the grid comes in.

The grid is the means by which we produce, store, and transmit energy. It is the massive and complex system of power stations of all kinds—coal- and gas-fired, nuclear power stations, wind turbines, solar power, energy derived from geothermal sources underground, wave power, and hydropower. It has to respond to daily and seasonal fluctuations and surges in demand, and in supply. The issue of grid resilience is therefore an important one to sort out.

Our means of storing renewable energy must improve, too. The traditional way to store renewable energy has been to convert it to hydroelectric power. In this way, spare energy is used to pump water into reservoirs that can generate electricity when needed. But building dams and using large quantities of water are not ideal environmental solutions.

Battery technology is getting better but needs investment, and is more of a solution for smaller-scale supply issues or for special cases, such as the massive facility built by Tesla in Jamestown, Australia. It is currently the world's largest battery, and can provide temporary backup power for 30,000 homes in South Australia. A bigger one, with more than 53,000 individual batteries, is planned for the Gemini Solar Project, a huge facility being built in the Nevada desert northeast of Las Vegas. We'll invest in battery technology, too.

Another way is to store the electricity chemically. Regular lithium-ion batteries can only hang on to their store for so long before it leaks away, but storing energy in a chemical bond can be more stable. Renewable methane is also promising. It's the method of storing electricity in a chemical battery by making methane from carbon dioxide; the methane can later be burned in a closed loop that retains the carbon dioxide.[21]

But hydrogen is perhaps the most attractive candidate. When renewable energy is creating a surplus of electricity, we can use it to make hydrogen gas through electrolysis, the splitting of water into its constituent parts of oxygen and hydrogen. The hydrogen is then stored until we need to burn it to make energy again, in a fuel cell that creates water as a by-product.

Hydrogen can also be used to heat our homes (replacing natural gas) and to make biofuels. It can power shipping and other heavy transport, including trains, buses, and trucks. And it can be used to drastically reduce the emissions of heavy industry that uses fossil fuels, such as cement and steel production. We'll make all this happen faster and at a scale that means the costs of hydrogen production and storage come down as fast and as far as have wind and solar. We'll aim to make it economic to use hydrogen in all of these areas of the economy. "It will be expensive to get to a hydrogen economy," says the carbon analyst and writer Chris Goodall, "but there isn't really any alternative. Renewables plus hydrogen may also be the cheapest way of getting to net zero."[22]

Support for a hydrogen economy has ebbed and flowed, the sticking point being mainly that almost all hydrogen produced at the moment is made directly from fossil fuels, mostly methane. A small amount of hydrogen is currently produced by electrolysis, but that process is itself powered by fossil fuels. What is needed is genuinely green hydrogen, where the electrolysis of water is powered by renewable energy, and we need this on a huge scale. The good news is that the technology is ready—the hydrogen-driven transformation of the economy is already underway, albeit slowly. In Sheffield, England, for example, ITM Power has opened the world's largest green hydrogen electrolyzing plant.

A modest (by our standards) investment in this area will help speed the transition. It is already as cheap to make hydrogen by electrolysis from renewable sources of energy as it is to make it conventionally, from fossil fuel.[23] And a 2020 report from BloombergNEF[24] showed that hydrogen made from renewable energy will become as cheap as natural gas, but obviously without the associated carbon emissions of the fossil fuel. The report finds that this clean hydrogen could cut 34 percent of greenhouse gas emissions, including from hard-to-decarbonize industrial sectors such as steel-making and shipping. Economics is already driving this transition; it won't need a huge investment from us to speed things along. As the technology matures, it will get cheaper, and better electrolysis will make it easier to store surplus grid energy, by driving hydrogen production.

Another improvement that will bring resilience and flexibility to the grid is if we can encourage trade between places that have power and those that don't. In Spain, all the wind power entering the grid is controlled by a single operator, Red Eléctrica de España, which means it can fine-tune supply and direct power to regions where the wind isn't blowing. This sort of model needs to be built into the national grids of all countries and regions: Different countries will find opportunities to trade their power when they have excess.

It will cost money, but that will quickly be recouped. The US Department of Energy produced a report in 2019 suggesting that a new energy transmission "super-grid" could be built to connect the entire contiguous United States for $80 billion. At the moment the Rocky Mountains divides grids in the east and west, and the new network would link them. The benefit would be huge, as surplus power could be delivered across the time zones, evening out the supply problems associated with renewable energy. And here's the killer point: The project is estimated to deliver economic gains of more than $160 billion.[25] The only reason this project is not (yet?) going ahead is that the Trump administration blocked it in order to protect the coal industry.[26]

MARK JACOBSON IS DIRECTOR of the Atmosphere/Energy program at Stanford University and an expert in renewable energy. He solves the problem of the grid by using multiple sources of renewable energy. In 2017, Jacobson and colleagues at the University of California, Berkeley, produced a groundbreaking roadmap for a global transition to carbon neutral energy by 2050. That's *all* forms of energy, not just electricity, but the harder stuff such as transportation, heating, cooling, and the energy used by industry. The team demonstrated how the grid will stay stable with a 100 percent transition to renewable energy in 139 countries.[27]

Jacobson's report was criticized for what some said were unwarranted assumptions about how surplus energy can be stored. His team have since tackled these concerns by dividing the 139 countries into 20 regions. In a simulation, the team checked grid stability in each region every 30 seconds for five years, and determined the resulting cost per unit of energy. "Many possible solutions to grid stability with 100 percent wind, water and solar power are possible," their new paper states. The report demonstrates, contrary to earlier conclusions, that the grid can stay stable even when supplied solely by renewable energy.[28]

Jacobson says there is no technical or economic barrier to transitioning the world to 100 percent renewable energy at low cost. By low cost he means the end price of energy, not the installation and the transformation of infrastructure needed. His study found that by 2050 the cost of renewable energy is 75 percent less than fossil fuels, mainly because we avoid the health costs associated with air pollution.

If there is no technical or economic barrier, how do we do it? The barrier is political, at every level. Some people pray that China is going to take the lead and hope this will drag other countries along. But, as we saw, China is currently installing enough coal-fired power stations at home and abroad (as part of its multinational infrastructure project, the Belt and Road Initiative) to replace *all* the carbon emissions of the European Union.

It might be easier to start changing things at a local level, and Jacobson is producing local roadmaps that can help make the transition at a smaller scale. This is where we can help, by starting small and snowballing. The demonstration of the practicality and economic benefits will drive greater uptake. But we do need to move fast on this. In the US and Europe, coal is dead, pretty much; it just needs finishing off. We have to invest in renewables to prevent natural gas or oil taking its place. One study found that, across ten Rust Belt states, from New Jersey and Maryland to Pennsylvania and Ohio, putting in renewable facilities to get electricity generation up to 13 percent by 2030, as required by state policy, will cost $3.5 billion. It will deliver $2.8 billion in savings from climate change impacts, *and* $4.7 billion in health benefits. Much of the power generation in those states at the moment comes from coal, and the resultant pollution causes a range of medical problems.[29] Groups such as America's Pledge—with signatories from cities, states, and businesses representing more than half of the country's emissions—are committed to meeting the Paris goals. We can make numerous windfall investments in renewables to help draw in investment and reverse the wasted time of the Trump administration.

Jacobson points to a range of areas that need immediate financial attention. First, of course, there is solar and wind generation, both on- and offshore. Then there are electric cars, and heat pumps for buildings (for both warming and cooling). Industrial energy needs are particularly hard to decarbonize. We need to replace fossil fuel-powered furnaces with electric arc or induction furnaces, or hydrogen alternatives. We need to provide better heat, cold, and electricity storage; expand the use of high-voltage direct current electric power transmission; and replace gas stoves with electric induction cooktops.

I asked Jacobson for reassurance that the $1 trillion would be enough. "Not really close," he said. "The roadmap would require on the order of $100 trillion in capital cost between now and 2050." At first, that knocked the wind out of my sails. But that's a global transition for *all* energy sectors, spread over 30 years (so not much more than $3 trillion a year). It includes electricity and transport, as well as all the world's needs for building, heating, and cooling; all industrial operations such as steel and cement production; and agriculture and fishing (which I've hived off into chapter 8). It also includes the military, which I've been ignoring.

Jacobson said "on the order" of $100 trillion but his analysis actually comes up with investment of $73 trillion (so, closer to $2 trillion a year), and he shows how this would pay for itself through both energy savings and the social costs of pollution on health and environment.[30] For the US, Jacobson's roadmap, which corresponds to the Green New Deal, requires construction of 288,000 large wind turbines and 16,000 large solar farms, creating 3.1 million jobs and needing initial investment of $7.8 trillion. Our pot of money can't get that whole job done, but it can start an unstoppable transition to net zero, one that has been going ahead in eleven US states already, even before President Biden reversed his predecessor's decision and rejoined the Paris agreement.

∽

WE'VE LOOKED AT ELECTRICITY GENERATION, and for that it makes sense to put the bulk of our money into wind and solar. But I want to veer off Jacobson's roadmap for a moment.

James Hansen is a former NASA climatologist, and the person who first raised the issue of climate change and its dangers in a testimony to the US government in 1988. Director of the Program on Climate Science, Awareness and Solutions at Columbia University, and an activist for action on climate change, Hansen was high on my list of people I wanted to consult for this chapter, and he was happy to give some advice on how to spend our money. Despite the obvious appeal of renewable energy, part of our money should go to nuclear power, he suggests.

It's an issue that divides environmentalists, due to nuclear accidents such as Chernobyl and Fukushima, as well as concerns over the future storage of nuclear waste. But coverage of these accidents has distorted understanding of the relative safety of nuclear power generation compared with fossil fuels. And the fact is that nuclear power provides a large proportion of carbon-free power. It doesn't make sense to retire working nuclear power stations prematurely. When this has happened, as in Germany since 2010, coal and gas have tended to fill the gap, adding to our carbon emissions.[31]

Another problem with nuclear is that the way we do it is desperately expensive. The cost of building Hinkley Point C, a new nuclear power station in Somerset, England, is now estimated at around $29 billion. When you compare that to the plunging cost of wind and solar, it is hard to justify. Hinkley electricity will cost about $123 per megawatt hour, while UK offshore wind farms are currently bidding to fulfil energy contracts at less than $53 per megawatt hour.[32] So, if we need nuclear power in order to kill off fossil fuels, we need to do it a different way. Not the massive and expensive reactors like at Hinkley Point, but far smaller ones, made in factories. This is what Hansen advocates, and many agree.

Modular nuclear power stations can be built at a central plant and delivered to the site to be assembled, a bit like an IKEA wardrobe. They are more efficient than conventional reactors because they burn much more of the nuclear fuel, so the waste problem is alleviated, too. They are safer and, it is claimed, immune to meltdown. "We have had the basic knowledge on how to do this for more than fifty years," Hansen says, "but the government chose not to invest in it." Now we can put this right. "Factory-built modular nuclear reactors could be cheaper than coal or other energy sources, and they certainly would have the smallest environmental footprint."

The International Atomic Energy Agency says there are about fifty designs and concepts for small modular reactors (SMRs), some of which, in Argentina, China, and Russia, are nearly built.[33] There is a nuclear engineering company in Portland, Oregon, NuScale Energy, with advanced plans for small modular reactors[34] projected to cost about $3 billion.[35] Its inaugural plant is slated to open in Utah in 2026. In the UK, the British National Nuclear Laboratory is "excited" about the prospect of SMRs and sees a large global market.[36] We'll invest in the company and accelerate production. The biggest sticking point for SMR is getting the production to a level where economies of scale kick in. A big order here for twenty or thirty reactors could overcome this hurdle. Timing is still a problem, as we need to decarbonize more quickly than nuclear reactors can be built and come on line, but a big sum here will speed the process.

⌐⌐

SMRs AND CONVENTIONAL NUCLEAR REACTORS work by nuclear fission, the splitting of radioactive elements to release energy. We will also look into speeding the construction of a working nuclear *fusion* reactor. Fusion is the way the Sun creates heat, and to do that in a controlled way on Earth means heating hydrogen to 100 million°C (180 million°F) until it forms a plasma, then getting the hydrogen atoms to fuse and produce helium. The act of fusion produces vast amounts of energy. You have to put a lot of energy in to heat it up, but the idea is you get ten times the amount back.

Steven Cowley is director of the US Department of Energy's Princeton Plasma Physics Laboratory. Like other proponents of nuclear, he says we need it to balance the natural fluctuations in energy delivered by wind and solar. "Fusion could do that easily once we have made it work," he says. That "once we have made it work" is the thing, though: For decades, fusion has been said to be twenty years away; it is one of those scientific and engineering challenges that is much, much harder in practice than in theory, certainly with the level of funding it has received in the past.

Teams working in France on the International Thermonuclear Experimental Reactor (ITER) have been trying to build a test reactor for years and they remain a long way away.

Cowley says he could make a large demonstration fusion reactor for $30 billion. To do so, he says, we could collaborate with the Chinese, who are working on their own fusion projects. The HL-2M machine, in Chengdu, became operational in December 2020. Then there are start-ups hoping to win the expensive but lucrative race to fusion, often by trying to make compact reactors. One, Commonwealth Fusion Systems, has investment from Bill Gates, Jeff Bezos, and India's richest man, Mukesh Ambani. UK Atomic Energy Authority signed a deal in 2020 with nine companies to develop fusion, including a plan to develop something called STEP, which stands for "Spherical Tokamak for Energy Production," by 2040. They say STEP would be the world's first compact fusion reactor.*

Cowley points out that, although $1 trillion is a huge sum, the global energy industry is worth $7 trillion annually. "Thus ultimately the goal must be to take it to the point that industry invests," he says. I'm willing to invest a small amount (relative to our total pot) in fusion, mainly because I'd like to see someone crack the development of small fusion reactors, and that's because you can see them in the future helping to power cities, spacecraft, and heavy industrial processes (preferably off-planet). But I can't see fusion being the solution we need, because we need one now.

❧

---

* Compact reactors are far cheaper to build, and new superconductors mean you can more easily make reactors with the magnetic fields necessary to contain the hydrogen plasma as it is heated to fusion temperatures. New developments in machine learning, a kind of artificial intelligence, mean it is also becoming easier to understand and contain the plasma.

We've seen that, using renewables for electricity generation, some of the hard work has already been done. The price has dropped so much as to make them competitive, and what we need to do now is roll out deployment. Much of our investment should be spent to this end. But this is far from job done.

In 2017, electricity generation contributed only 26 percent of global greenhouse gas emissions. We rely on fossil fuels and greenhouse gases for a huge range of other things, and all of these need to be decarbonized, too.

The biggest single sector that we could change for the greatest reduction in greenhouse gas emission is refrigeration. In the past, air-conditioning units as well as refrigerators used chlorofluoro-carbons (CFCs) as the coolant gas. CFCs were banned following the discovery that they were destroying the ozone layer. In a swift and decisive move, the Montreal Protocol on Substances that De-plete the Ozone Layer prohibited their use. (One can only imag-ine what might have been if 1997's Kyoto Protocol to limit carbon dioxide emissions had had a similar effect.) However, despite the Montreal Protocol, many older devices that contain CFCs re-main, and when they are scrapped the gas escapes into the atmo-sphere unless carefully removed. CFCs, and their replacements as coolants, hydrofluorocarbons, have far higher warming poten-tial per molecule than carbon dioxide—they are many thousands of times more potent. The chemicals are being gradually phased out but it will be years until all fridges and air-conditioning units containing them are decommissioned. We need to spend some money on speeding the transition to alternative coolants such as propane and ammonia.

Our buildings, too, need urgent attention. They are responsible for 23 percent of carbon emissions. Many aren't well insulated, and the heating and cooling systems we use either derive from fossil fuels or from other even more powerful greenhouse gases, as we've just seen. So we need to retrofit our homes and offices and shops to go carbon zero, with better insulation, double glazing, green roofing, new lighting (all our lighting needs to change from old-fashioned bulbs to LEDs). This is an expensive initial outlay—beyond our means alone—but doable, with government support, and pays back on the energy savings that will accrue. In 2019, New York City completed its replacement of all the windows in the Empire State Building, as well as all the heating, cooling, and lighting, saving 38 percent of energy use.

Diana Ürge-Vorsat, a Hungarian climate scientist and an IPCC vice chair, says that to her mind there is absolutely no question that deep retrofit is where we should invest our money. It's a subject that we don't hear much about, because retrofitting spans multiple industries and doesn't offer an easily grasped opportunity. We need to create, she says, an industry body to lobby and organize retrofitting at a high level, a one-stop shop where city, state, and national governments can get advice from architects and get retrained.

We also need to make all new construction carbon neutral. As we'll see in a moment, this will mean changing the kinds of materials we need for construction. Again, it can be done. In California, all new residential buildings now have to be made to consume zero net energy. All existing buildings in Cambridge, Massachusetts, have to be net zero by 2040.

AS THE WORLD'S FIRST TRILLIONAIRE, I was looking forward to buying myself a zero emission private jet, to help me visit the different projects I'm funding. Imagine my disappointment in discovering that no such aircraft exists. Or not yet, anyway.

Aviation accounts for 2.5 percent of global greenhouse gas emissions. That doesn't seem much, but its real effect may be twice that number, as the emissions take place at high altitude.

Getting rid of this source of carbon will take some effort, because, unlike road or rail vehicles, you can't fly aircraft very far on battery power. Solutions are being sought. NASA is developing an electric airplane, the X-57 Maxwell, and major aircraft manufacturers are working on efficiency savings and design improvements, as well as new kinds of engines and fuels. Rolls-Royce, for example, is working on a hybrid electric/kerosene engine, while others are developing SAF, sustainable aircraft fuel made from recycling waste products.

ZeroAvia, an aircraft company based in the US and the UK, has started test-flying a prototype electric plane that seats six passengers plus crew. At the moment a battery-powered aircraft looks good only for small planes and short commuter flights, so the start-up is developing a hydrogen fuel cell to power its aircraft. In September 2020, ZeroAvia flew a six-seater running on hydrogen instead of kerosene. It was the first hydrogen-powered flight of a commercial-size aircraft. Eventually you'll be able to produce electricity in flight from hydrogen, making heat and water as waste. However, compressing or liquefying hydrogen is currently very expensive. We'll give this a kick.

Electric vehicles, on the other hand, are rapidly maturing. Batteries are starting to work well for cars, and are becoming competitive in price and performance. Police departments in the US, for example, are starting to use Teslas because they are more efficient and cheaper to run than petrol/diesel models (and have better acceleration). Electric cars get a lot of publicity, especially when they get launched into orbit around the Sun as part of a tie-in with a sibling company (as happened when SpaceX launched Tesla), but currently make up only a tiny proportion of the 1.2 billion personal vehicles in use.

While we start to get petrol-based cars off the road, we need to improve their efficiency; we also need to increasingly switch to electric vehicles for personal use and especially buses and trucks.

But, most of all, we need to encourage substitutions: Get people to use public transportation, or to walk and cycle instead. The carbon footprint of an electric car—the amount of carbon emissions created in producing one—is higher than a conventional vehicle, thanks mainly to the exotic elements required for the battery. We don't really want to replace a billion petrol-powered cars with a billion electric ones.

Electric vehicle manufacturers such as Tesla and Nissan are doing well, so our investment in this sector should be modest, mostly to help speed the transition. An improved infrastructure for trains, and overall a modern, self-charging smart network for public transportation, would have a greater impact—but you see how our trillion dollars soon gets eaten up.

WE SHOULD ALSO INVEST in carbon-neutral synthetic fuels. These fall into two categories: biofuels, which are fuels made from plant material, and electrofuels, which are produced from carbon dioxide and electricity.

A common way to make biofuels has been to take a crop plant, such as sugar cane, and ferment it down to make ethanol. This can then be refined into a fuel for burning in an engine. The obvious problem is that to do this we have to give up land we need to grow crops for animal and human consumption. We don't have that land to spare (see chapters 4 and 8).

Electrofuels seem to be more promising. The US-based Advanced Research Projects Agency–Energy (ARPA-E) has a well-developed electrofuels program using microorganisms. If transportation fuels can be made cheaper than oil-derived fuels, we're laughing. You could take hydrogen derived from a renewable source and use it to make hydrocarbon fuels. You could also use bacteria that digest ammonia and convert carbon dioxide into liquid fuel.

Other teams have modified *E. coli* bacteria to metabolize carbon dioxide itself, or get bacteria that harness solar power and use this to create the fuel. There are many such projects, but we can't yet make carbon-neutral fuel at scale, not at a price that is competitive.[37] We should invest in this area as another route to decarbonizing the aviation and shipping industries, and because a recent projection has synthetic fuels as being ultimately more competitive than batteries for cars driving long ranges.[38] The authors recommend developing various forms of synthetic fuels in a Darwinian manner, to discover the best route to scaling up production and achieving decarbonization.

Eventually, shipping, which composes 3 percent of total global emissions, will be powered by synthetic fuels, or hydrogen, or perhaps by solar. "Slow steaming" may be one of the ways to cut emissions before we have synthetic fuel production at the necessary scale. Reducing the speed of ships by 20 percent cuts emissions of carbon dioxide, as well as emissions of sulfur and nitrogen oxides.[39]

Half to two thirds of all the oil used in the world goes into fueling our cars, trucks, buses, trains, ships, and planes. Transportation accounts for one quarter of all energy-related carbon dioxide emissions and is growing.[40] As with all our projects in this chapter, there are great savings to be made from moving to sustainably powered systems.

⌒⌒

IT'S GOING TO BE HARD to get to zero-carbon transportation. It will be harder still to make the products of heavy industry—cement and steel, in particular, but also aluminum, plastics, and ammonia—using renewable sources of energy. Concrete and steel together make up half of the emissions from the industrial sector. Jeff Bezos has talked of moving heavy industry off-planet, and you can see why. While we wait for that to happen, we need to decarbonize production on Earth.

Concrete is made from sand, rock, water, and cement in a desperately carbon-heavy process. Cement is made by heating limestone in kilns at high temperatures, which takes a lot of energy, and which makes carbon dioxide as a by-product. Some can be captured, and we can use electric kilns, but nothing is approaching scalable levels or is anywhere near economically viable yet.

Steel is just as bad. We made two billion tons of it in 2018. Again it takes a lot of energy to get a furnace hot enough to extract iron from iron ore and convert it to steel, which itself produces carbon dioxide as waste. We can replace conventional furnaces with electric arc furnaces, and try to use hydrogen to make the steel, which would produce water as a by-product rather than carbon dioxide, but this is only at the prototype stage as yet. Both these processes—steel and concrete—could burn up a fair bit of our cash if we're not careful.

∽

THERE IS NO MORE PRESSING MISSION facing humanity than avoiding catastrophic climate change, by which I mean avoiding more than 2°C (3.6°F) of warming. In 2009, the International Energy Agency (IEA) said that each year the world delayed making major cuts to emissions will cost an extra $500 billion. In 2011, the IEA said that for every $1 investment in clean technology that is avoided before 2020, an additional $4.30 will need to be spent to compensate. The Paris Agreement committed signatories to NDCs—Nationally Determined Contributions—which last until 2025 or 2030. The Glasgow Climate Pact of 2021 committed all countries to ratcheting up those NDCs; if they don't, we could see warming of around 3°C (5.4°F) by 2100, with the associated crises of species extinctions, food insecurity, sea level rise, drought, and so on. The UN Environment Programme said that we currently need to cut emissions by 7.6 percent each year until 2030 to get off this trajectory. Despite our efforts so far, and despite the COVID-19 global lockdown, we've never managed to reduce overall annual emissions.

To me it's looking like the bulk of our money is going to wind and solar power. We should certainly give hydrogen power a boost, and spend a small amount on research into areas such as electrofuels and nuclear fusion. We can't hope to get close to renewing global infrastructure or heavy industry, or manufacturing. Nor should we expect to: After all, the task is to change a global civilization that has based itself on fossil fuels for more than two hundred years. Instead our investment will go toward the deployment of renewable energy systems as fast as possible. As Jacobson says, "Deployment itself lowers costs and spurs competition, creating its own research."

Under the terms of $1 trillion bequest, we're prohibited from directly lobbying governments or engaging in political activity. But a final word about subsidies. An International Monetary Fund report found that subsidies paid to fossil fuels dwarf the levels per year that we have to spend.[41] In 2015, the US alone spent $649 billion on direct and indirect subsidies for coal, oil, and gas ($50 billion more than it spent on the military); the global spend was $4.7 trillion in 2015 and $5.2 trillion in 2017. Fossil fuel subsidies are calculated as the difference between consumer fuel prices and what the price would be if the true supply cost was taken into account, plus the environmental costs of pollution, including the impact on human health. The report found that removing subsidies would cut global emissions by about 25 percent, and the number of premature deaths from air pollution by about 50 percent.

Achieved

Acceleration of the transition to net zero with investment in a range of schemes and industries; the preservation of a world both prosperous and sustainable, livable, and equitable.

Money spent

Deployment of renewable energy capacity:	$860 billion
Development of the hydrogen economy (including electrofuels, biofuels):	$50 billion
Incentives for low-carbon rail infrastructure:	$20 billion
Incentives for electric vehicles:	$5 billion
Development of nuclear fusion:	$5 billion
Development of modular nuclear power:	$30 billion
Incentives for carbon-zero buildings:	$10 billion
Decarbonization of industrial processes:	$20 billion
Total:	**$1 trillion**

A naked mole rat—surely one of
the greatest mammals of Africa.

4

Save Life on Earth

AIM: To halt the human-driven erosion of life on Earth in order to maintain the habitability of the planet and protect biodiversity. To save endangered species and catalog their genetic material. To protect rainforests, wetlands, coral reefs, and all areas of key biodiversity value. To reduce the current catastrophic rate of species extinction while at the same time reframing our relationship with the natural world.

THE NAKED MOLE RAT is spectacularly ugly. Hairless and wrinkled, it has a giant pair of incisors growing through its lips. It is not an obvious choice to start a chapter celebrating the wonder and diversity of life on Earth, but it truly is one of the most fascinating of all mammals. It lives underground, digging tunnels with its teeth, which is why the teeth grow through the lips: it can keep its mouth shut while biting through the soil. For a small rodent, it is exceptionally long-lived, getting to thirty years or more. If humans lived the same time relative to body size, we'd live to six hundred. They're also immune to cancer and pain, and are the only nonhuman mammal to bury their dead. And then there's their way of life, which is more like a social insect, in a colony ruled by a queen. I said they were ugly, but I have grown to love them.

The world is full of life, and full of different kinds of life. From a bacterium that uses the radioactive decay of uranium as its sole source of energy to the swift, an animal that feeds, mates, and even sleeps on the wing, whose very habitat is the air. How many species there are on Earth, we just don't know. It's a problem we'll come back to, and a serious one, because we won't know how much land and sea we should protect if we don't know how many species there are and how much area each one needs, but at a rough estimate ecologists say there are about ten million different species on the planet.

"Biodiversity" is the umbrella word used to encapsulate all of this. It's such a buzzword these days that it's strange to think that it hasn't been around all that long.* Perhaps the richness and variety of life just wasn't a concern for most of human history, because there was so much of it. Certainly we don't seem to stop hunting and fishing until the species we're after has gone extinct. I wonder what prehistoric people thought as they discovered that woolly mammoths were becoming harder to find. Did any Paleolithic genius consider that they'd overhunted the animals? I don't imagine anyone at the time stopped to consider the demise of saber-toothed tigers in any terms other than relief, and the same with American dire wolves, which now are found only in museums and on *Game of Thrones*.

Extinction simply hasn't been much of an issue for most of the Holocene, the interglacial era we're in at the moment, which has lasted 11,000 years. But the rate of species extinction rate is now

---

* The word was first used by wildlife scientist and conservationist Raymond F. Dasmann in 1968, and was widely adopted only in the 1980s.

far higher than the "natural" rate, with estimates ranging from hundreds to thousands of times higher.[1] For some groups of animals, the rate is even worse: Amphibians are going extinct 25,000 to 45,000 times higher than "natural."[2] It's why some biologists think that a mass extinction, the sixth[3] since life evolved (or maybe only the fifth),[4] is underway. Whether or not the dying off that is happening now qualifies is up for debate. The biggest mass extinction, at the end of the Permian 252 million years ago, wiped out 90 percent of all species; the asteroid that did in the dinosaurs, 66 million years ago, took out 75 percent of all species. We are nowhere near those figures, but the die-off is *bad*. We are seriously endangering the functioning of our planetary life support.

FOR A MOMENT, let's stick with the big charismatic things we care most about. There are some 5,416 known species of mammal: from kangaroos and koalas to elephants and blue whales, from dugongs to Kodiak bears and jaguars. Our closest relatives are, of course, the primates—the monkeys and apes—among which 60 percent of the 504 species are predicted to go extinct in the next twenty to fifty years. It feels wrong to rank species in terms of their emotional value, but it will be a moment of scarcely endurable shame if—when—a great ape species goes extinct in the wild. It seems only a matter of time, since all thirty-nine species and subspecies of ape are officially threatened with extinction.[5] In total, a *quarter* of all mammals are so threatened. That figure is 13 percent for birds, which amounts to 1,375 species.

Arguably more worrying, however, is the decline of insect species and populations, which has been so great as to have been called, with no little hype, an apocalypse. In 2019, a review of seventy-three studies of insect diversity found that more than 40 percent of insect species were threatened with extinction.[6]

Entomologist George McGavin of the University of Oxford told me that the most shocking thing he'd seen in his career was children hand-pollinating pear trees in China. The apple and pear orchards in southwest China produce high-value fruit, but excessive use of pesticide has wiped out the bees and other insects that pollinate the flowers, so farmers have to carry paintbrushes and pots of pollen and hand-pollinate each flower. Children are used to climb to the higher branches. It's just about economically feasible with pears and apples, and with a low-cost workforce, but it wouldn't work for the rest of our crops.

If the insect collapse that has hit China hit the rest of the world, we couldn't fill in by hand-pollinating everything: raspberries, strawberries, peas and beans, zucchini, tomatoes, and blueberries. Or perhaps we could, on a tiny scale, as a luxury item that only the 1 percent could afford. The rest of us would have to make do with wind-pollinated wheat, barley, and corn. That's not to mention the other jobs insects and other invertebrates do, forming the soil and breaking up and recycling dead animals and plants and their waste. We don't know the full extent of the collapse in insect numbers, but a 2014 paper felt confident enough to pronounce an "Anthropocene defaunation."[7] The uncertainty highlights the lack of knowledge and monitoring that we have of the planet. This is something we will change with some investment.

THE EXTINCTION CRISIS is not on the scale of the apocalyptic episodes in the deep past; it doesn't seem to be a mass extinction in a geological sense. But that doesn't matter: If we don't get a grip on it, it could have a profound effect on life on Earth, and that includes human life. Our job in this chapter is to turn the crisis around. First, however, we must digress to find out why diversity is necessary.

People might say—indeed, people do say, as if manifest destiny was still a thing that could be talked about in civilized company—"we're the ruling species on this planet, it's tough luck for those that get in our way." In 2015, Andy Conrad, CEO of Alphabet's Verily, said, by way of explaining the name of the company: "Only through the truth are we going to defeat Mother Nature."[8] Logging supply companies sell T-shirts and bumper stickers that read, with a phrase that would be funny if it wasn't so chilling: EARTH FIRST! WE'LL LOG THE OTHER PLANETS LATER.

The problem is not that logging companies believe we should strip bare the entire land surface. It's that it's their way of life and they have not been offered an alternative. The same is true for people in any area that we might want to protect, and it's a problem that traditional conservation methods have often neglected or ignored, and one we'll try to keep front and center of our approach in this chapter: Conservation has to take into account people's needs.

Ultimately, people's needs coincide with those of other lifeforms. The biosphere is literally the life support system for our rock traveling through space. It provides clean water and produces oxygen; it cycles the water around the planet; it regulates the climate and cleans the air. If something goes wrong on our spacecraft, we will get into trouble. Remember Apollo 13 and "Houston, we have a problem?"* Something went wrong on the way to the Moon and carbon dioxide in the spacecraft started building up to dangerous levels. The analogy to the world right now is uncomfortably close, but where it breaks down is that NASA was able to save the lives of the three astronauts on Apollo 13 through engineering ingenuity. Our life support problem is on another scale.

---

* "Houston, we have a problem" is a mutated meme. The original phrase, said by astronaut Jack Swigert, was: "OK, Houston, we've had a problem here."

A rich biosphere is essential for ecosystems to function properly. When it is healthy, it runs the life-support systems we rely on—things such as the production of oxygen and clean water, pollination, the cycling of nutrients that underpin life, coastal protection, provision of timber, commercial fisheries, and carbon storage—it provides these things so invisibly that we've taken them for granted. Ecologists have called these functions ecosystem services, but some have moved away from this and now prefer to say "nature's contribution to people" (NCP).[9]

Framing the relationship between people and nature as a stock-and-flow system risks commodifying it, and alienates many cultures; the NCP approach is more inclusive. In general, a more biodiverse ecosystem is better at providing these contributions, is more resistant to invasion by weeds and "alien" species, and is more stable.[10] The greater the diversity of an ecosystem, the greater its productivity and its potential to lock up carbon dioxide. Even if we didn't care about gorillas and whales and rainforests and coral reefs for their own sakes, we should care about maintaining our life support system.

We also should care because it will cost us money if we don't look after it. Huge, economy-crippling amounts. One study estimated that the financial loss of ecosystem services that can be attributed to changes in land use from 1997 to 2011 amounted to between \$4.3 and \$20.2 trillion per *year* (with the massive variation due to uncertainty in the precise value of each type of habitat).[11] World GDP was \$46.3 trillion in 1997 and \$75.2 trillion in 2011, so you get an idea of the scale of the problem.

There are several areas to focus on. We need to identify the key factors driving the extinction crisis and find ways to stop them. We also need to identify the key areas on the planet that need safeguarding.

First, the causes. The acronym "HIPPO" captures the main ones: Habitat loss, Invasion by non-native species, Pollution, Population increase of humans, Overhunting. Climate change is perhaps the most important element in the first three factors, and indeed climate change and biodiversity are intimately linked. In 1987, at a meeting to plan for the Convention on Biological Diversity, the ecologist Thomas Lovejoy remarked that, if we didn't tackle climate change, then we "could forget about biological diversity." (This was in the years before the phrase was contracted to "biodiversity.")

In 1990, John Harte and Rebecca Shaw, then of the University of California, Berkeley, started the world's first climate change ecology experiment.[12] Their aim was simple: Instead of making computer models and predictions of what might happen, they would warm a piece of land and watch what happens. They chose a montane meadow ecosystem in the Rocky Mountains, Colorado, and set up plots of land that were experimentally warmed.

What happened, as the years of experimental warming went by, mimicking what we can expect in the coming years, was that the plant and soil community structure changed. Sagebrush started to dominate, and crowded out the meadow species, the grasses, and flowers. This didn't help the insects and birds that fed on them. In a nasty double whammy, Harte and Shaw found that the change also resulted in a substantial loss in the ability of the soil to store carbon.[13] As sagebrush takes over, more carbon dioxide is released into the atmosphere.

Climate change and biodiversity are intimately bound up. But biodiversity has a chapter of its own in this book because it is so vital to our survival. A diverse planet is a stabilizing planet and a buffeting planet; it can roll with the punches.

The biggest HIPPO in the list is habitat destruction. Just looking at forests, the planet is losing 29,000 square miles per year, which is twenty-seven soccer fields of forest per *minute*. Some

17 percent of the Amazon has been lost in the last fifty years. At the turn of the century, tropical forests covered between 10 and 15 percent of the planet's terrestrial surface; today that figure is less than 5 percent.[14] Satellite data showed that between August 2018 and July 2019, the Amazon lost almost 4,000 square miles of forest,[15] a strong upward trend in the rate of loss linked to the election of Brazilian president Jair Bolsonaro and his ecocidal promise to "develop" the Amazon.

The Amazon is the greatest of Earthly treasures. By far our largest expanse of forest, home to thirty million people, it influences the planet's climate through air and water circulation, and stores carbon equivalent to a decade's worth of human emissions. It creates its own rain, but this system fails if the forest becomes too small, as when it is converted to pasture or burns in wildfires. Positive feedback means the forest dries out, and is more vulnerable to fire, which destroys more of the forest. There is a tipping point, estimated at between 20 percent and 40 percent of forest cover, where fire and drying run away with the forest— and remember that we've lost 17 percent already.[16] If the Amazon goes, we lose an incomparable assemblage of life, and could face unmanageable levels of global heating.

Much of the deforestation is to clear land to raise cattle, to feed the insatiable demand for meat, as we'll see in chapter 8. It comes at a huge price, because forests sustain the largest numbers of species—they are the most biodiverse ecosystems on the planet, certainly on the land—and, as they decrease in size, so does their ability to support other species. Project Drawdown, the resource that assesses solutions for climate change problems, ranks deforestation as the single most important issue facing the global environment. So, forestry protection and restoration and expansion are going to be central to our spending in this chapter.

There is good news: Large amounts of carbon can be drawn down into forests and soils when they are allowed to naturally regenerate, as opposed to when areas of land are actively planted with trees. It's cheaper, quicker, produces a more diverse ecosystem, and is often more effective at capturing carbon dioxide to let forests regenerate on their own. A global study in 2020 found 2.6 million square miles, which could just be set aside and allowed to regrow, leading to the capture, by 2050, of 80 billion tons of carbon.[17]

In 2014, the UN climate summit saw the launch of the New York Declaration on Forests (NYDF), aimed at halving deforestation by 2020 and stopping it entirely by 2030. It's a stirring and worthy project to get behind, but there is an unhappy clue in the name—it's a declaration, not a treaty; it has no legal teeth. After the NYDF was launched, deforestation didn't stop or even slow down, it *accelerated*. It's like the surge in gun sales in the United States after each massacre, when people think it's the last chance to buy guns before the government clamps down on sales. Each year since 2014, the loss of carbon as forests are cleared has been equivalent, on average, to the greenhouse gas emissions of the European Union. A report by NYDF Assessment Partners shows the most intensive clearance is going on in the Amazon, in a region overlapping Bolivia, Brazil, Colombia, and Peru. But not just there. Deforestation in the Democratic Republic of Congo (DRC) doubled between 2014 and 2019.

The problem, of course, is that the short-term rewards to individuals for participating in illegal logging, and for the outfits that organize it, are too tempting, and outweigh the risks of getting caught. I understand the actions of individual members of logging gangs or miners in Brazil or DRC or Colombia or Borneo or Indonesia: They need an income like the rest of us. As Brazilian miner Jorge Silva put it: "All of us here realize we're fucking the environment. It's not like we want to—it's that we haven't found any alternative means to survive."[18]

Silva and the rest of us are being exploited by the prospectors and logging cartels that organize them and the governments that allow it. Their profit is coming at a terrible cost to the planet, to its ability to function, to all species that live on it. Let's not tilt into gloom. Let's channel this instead into fury. The NYDF as it is currently funded cannot hope to stand up to the forces of deforestation, so we will absorb it into an umbrella organization called Canopy and give it some muscle to the tune of $10 billion. This is the initial and immediate payment, but we will build in a scale of payments rising to $100 billion as we identify and engage in active forestry protection operations.

We will, of course, need to work with governments. Brazil's Environment minister, Ricardo Salles, said in 2019 that rich countries should pay Brazil to stop destroying the Amazon.[19] If farmers and local people across one fifth the area of the Amazon were paid not to develop their land through deforestation, it would cost us $36,000 per square mile a year, which comes in at $12 billion annually. The big problem here is not so much that we'd be funding the Bolsonaro regime as that we'd need assurance and proof that the Amazon really would be protected: Brazil needs to properly enforce the legal protection of the forest.

Much of our forestry investment will be spent on buying logging and mining concessions from national governments, and then protecting the areas. Crucial to the entire operation is to work with local people, and to ensure that any action benefits them over both the short and the long term. We will work with people dependent on the forests for their income and, crucially, with indigenous peoples whose traditional lands are at risk. They are often the most effective of forest custodians.

Preventing deforestation, and actively enhancing forest cover, will help both biodiversity and carbon management. We'll look at this second aspect, how tree planting and ecosystem restoration can mitigate against climate change by drawing carbon dioxide out of the atmosphere, in chapter 7. We also have to ensure we meet the nutritional and energy needs of a growing human population and will look at food supply in chapter 8. Our focus here is biodiversity.

∽

BEFORE I FIRST WENT DIVING, a zoologist friend told me I'd have to rethink everything I thought I knew about animal behavior and communication and ecology when I was underwater. And so it turned out. On a reef there are thousands of organisms interacting in different ways—feeding, hunting, growing, absorbing, mating—communicating unknowable information, all through a medium that we didn't evolve in and can't experience. As interlopers, we can barely grasp what's going on.

But that's part of what makes diving so special. Getting a glimpse of an alien world. The other big thing is just the ability to move effortlessly in three dimensions, to spin and rotate and orient yourself in any plane, hardly bound by gravity. It's as close as most of us will get to being an astronaut.

I learned to dive on a live-aboard boat on the Great Barrier Reef, the 1,400-mile wonder of the oceans, the world's largest continuous coral reef, and one of the most biodiverse ecosystems on the planet. Ove Hoegh-Guldberg, director of the Global Change Institute in Queensland, calls coral reefs *megadiverse* ecosystems. On a live-aboard you can tour different parts of the reef and dive throughout the day and night, so I got to see a good variety of coral types and got some small grasp on the teeming richness of the life they support. I remember on a night dive seeing the animals sleeping, including a parrot fish tucked up in a protective bag of mucus it had secreted. But the coral—the coral was transformed. At night is when coral feeds.

Coral is a symbiotic organism made of an animal part—a simple colonial animal related to jellyfish, which builds the hard reefs and the calcareous structure you see when you see coral—and a colorful photosynthesizing microorganism called a zooxanthella. The microbe converts sunlight to energy, which the animal uses to build the structure for the microbe to shelter in. At night the polyps stretch out of the coral to catch planktonic prey.

When I dived on the reef in the 1990s, the big concern for the coral was its vulnerability to the crown-of-thorns starfish, which was invading and eating the coral and which we'd see on our dives and avoid touching (a woman on a dive with me accidentally landed on one and spiked her leg). How quaint an ecological fear that seems now, and how manageable, in the face of the existential threat of climate change.

The oceans have absorbed 93 percent of the extra heat that has been generated by global warming, and about a third of the carbon dioxide we've emitted.[20]

The oceans are vast, so they have been able to absorb a lot. But we've reached the limit. Carbon dioxide dissolves in water and makes carbonic acid, so the oceans have been getting both warmer and more acidic. As this happens, the colorful zooxanthellae are forced out of the corals in a process known as bleaching. Neither organism can survive alone and, if the zooxanthellae can't recolonize, the reef dies.

Diving on the Great Barrier Reef now is to experience the unfolding tragedy of warming and acidification. In 2016 the surface waters of the reef warmed by between 1° and 3°C (roughly 3.5° to 5.5°F), and mass bleaching took hold, occurring over 90 percent of some reefs. If we keep global warming to 1.5°C (2.7°F), we can keep some of the corals—maybe about half of them. But at 2°C (3.6°F) we will lose 99 percent of coral worldwide.[21] Corals support about a quarter of all marine species, and provide ecosystem services to around 850 million people who live near them and benefit from their protection and food supply.

This is a particularly difficult problem for us, because the essential and urgent action is to prevent warming getting to 2°C (3.6°F), and reduce the acidity of the ocean. The problem of saving the coral illustrates why we have to think globally and change the way we live entirely.

What we *can* do with our money is to prevent mechanical destruction of the remaining reefs, and stop water pollution and overfishing at reefs. We can support efforts to preserve coral. We will fund the Great Barrier Reef Legacy, an organization planning to create the Living Coral Biobank in Queensland. It's a facility that will house and nurture eight hundred different types of living coral. We can regrow coral in the wild, too: Electrified metal frames submerged on reefs have been used in Southeast Asia and the Indian Ocean to accelerate regrowth of damaged reefs. Returning ranch land in Queensland to its former wetland state provides a buffer and helps stop the run-off of fertilizer that poisons the Great Barrier Reef. We must support people in coastal areas to give coral reefs every chance possible while we try at a higher level to slow global emissions.

⌁

WE HAVE NO REAL IDEA how many species we share the planet with. For biologists, this is our dark matter, the thing that we really *ought* to know about. The problem isn't quite as bad as the one physicists have with dark matter (as we'll see in chapter 9), because the species are out there and we can in principle go and find them. It's just there are so many, we haven't gotten around to it.

We've been trying in earnest since 1758, when the Swedish botanist Carl Linnaeus devised the system of species classification that we still use today, assigning each life-form a Latinized genus and a species name, both written in italics. The sparrow-hawk is *Accipiter nisus* and the stag beetle is *Lucanus cervus*. It works very well, but it takes time to carefully describe and assign new specimens, as species needing names come in faster than we can classify them.

Biologists have begun to loosen the rules over using morphological characteristics in the name—the evolutionary biologist Edward O. Wilson has said he ran out of Latin words to describe new ant species. So a beetle has been named for Greta Thunberg, *Nelloptodes gretae*, and a particularly striking octopus was given the name *Wonderpus photogenicus*. The uranium-powered bacterium is *Candidatus Desulforudis audaxviator*, the species name coming from the Latin message in Jules Verne's *Journey to the Center of the Earth*: "*Descende, audax viator, et terrestre centrum attinges.*" In English, "Descend, bold traveler, and you will attain the center of the Earth."[22]

One way or another, we've formally described and named just over two million species in the 250-odd years since Linnaeus. As for how many species there might still be out there, waiting to be discovered, one estimate puts the total number of eukaryotes at around 8.7 million species. (Eukaryotes are the "advanced" life-forms of animals, plants, and fungi.) Their estimated number on the land is 8.7 million; in the ocean the figure is 2.2 million.[23] That means we've still got more than 80 percent of the job of classifying species ahead of us. It may well be more than that, and we haven't counted microorganisms yet. Here's where things get really out of hand.

Life-forms tend to follow a pattern, whether they are plants or animals or microscopic organisms: There are more small things, and fewer big things; at a medium size there are medium numbers of things. This amazingly woolly, slightly Rumsfeldian statement can be recast mathematically and turns out to hold true across thirty orders of magnitude, and can be used to predict the total number of species at each level of scale. If you do this and look at the number of expected types of microorganisms, you get a prediction of more than 1 *trillion* species.[24] Sure, the whole concept of what a species even is becomes confused when you get to bacteria and the other massive group of microorganisms, the archaea, but you get the idea: There is clearly a lot of work to do. It's practically endless.

The problem is that classifying new species in the midst of a mass extinction event is like bailing water out of a sinking boat. It's why when new species are discovered, often they are immediately entered on the Red List of Threatened Species. The first new species of bird to be described on continental North America in a century, the Gunnison sage grouse, got slapped with this unwanted label. It's rare for a new mammal or primate to be discovered these days, and rarer still for a great ape; when a third species of orangutan was discovered in a fragment of forest on Sumatra in 2017, it was immediately declared critically endangered. In fact, the Tapnuli orangutan, the first great ape discovered since the bonobo in 1929, straightaway became the most extinction-prone of all great apes.[25] There are three species of orangutan, the only great ape found in Asia. They are less well studied than the African apes, but recent work shows they are no less intelligent, using tools to get food and capable of flexible, value-based decision making.[26] In Borneo, there are around 60,000 *Pongo pygmaeus* (they are being killed at a rate of more than 3,000 per year); in Sumatra there are some 15,000 *Pongo abelii* and just eight hundred of the new species, *Pongo tapanuliensis*.

Their plight is a clear illustration of needing to make conservation work for people's benefit. The habitat of the Tapnuli orangutan is threatened by the construction of roads and a dam, and not all of its forest habitat is under government protection.[27] There are two gold mines in the area. If you ask local people if they would rather have better infrastructure, a source of hydroelectric power, and perhaps some revenue from opening up the gold mines, or if they would relinquish this for the sake of the apes they've sometimes seen raiding their gardens, you may not get the answer you want.

So you see the catch-22 situation. We have to catalog what's out there so we know the scale of the task facing us: We need to know how many species there are in order to work out how much land they need to survive. At the same time, we're clearing the habitat where these species live before we even know what's there. Any conservation action must be with the permission and informed consent of people living in the area, and carried out to their benefit. One thing on our side is that conservation strategies can promote tourism, and provide a sustainable income for the area, whereas forest clearance for palm oil plantation, or for exploitation of rare metals, may not be in the local people's best long-term interest.

There are good examples of how people-centered conservation can be achieved. The Keo Seima Wildlife Sanctuary covers nearly 1,200 square miles of forest in eastern Cambodia.[28] It is a globally important area for biodiversity, with more than 950 recorded species, including dozens of threatened plants and animals. It is also home to the Bunong people. The area is managed by the government's environment ministry, with support from the Wildlife Conservation Society and other partners, who have worked to engage the communities with conservation measures designed to improve their livelihoods. For example, twenty villages in the area have signed up to a carbon credit scheme whereby they earn revenue by helping to keep the forest intact as part of the REDD+ project (reduced emissions from avoided deforestation and forest degradation). In 2016, the villages sold their first REDD+ carbon credits to a US company, using the revenue to improve their water supply and construct a meeting space for the community. It is by no means easy to manage, however. In 2018, three rangers were murdered while investigating an illegal logging camp.[29]

Another example is Madidi National Park in Bolivia, a 7,500-square-mile area of jungle with more than 20,000 plant species and 1,100 species of bird. A new species of primate, the Madidi titi monkey, was discovered there in 2006.[30] Established in 1995, the region is home to forty-six indigenous communities from six different tribes. Tourism is the main source of income and is used to help local people live sustainably and improve their quality of life. As with Keo Seima, the area is supported by both government and NGO sources.

So, it can be done. The Wildlife Conservation Society (founded in 1895) works with governments around the world on similar projects. We will create an organization to oversee them, and expand the muscle and reach of existing NGOs.

~

WE ALSO NEED SAMPLES from species so we can preserve them. It is too late to save many species, but we can at least record their DNA. Perhaps in the future we will be able to resurrect extinct animals.

The collection of varieties of crops for breeding was first done on a large scale in the Soviet Union in the 1920s. In the United States, the national gene bank was established in 1954. The International Treaty on Plant Genetic Resources for Food and Agriculture attempts to ensure that global crop diversity is properly documented and preserved. The Svalbard Global Seed Vault was set up as an international facility to store seed crops, and now contains more than a million different varieties. Its location, inside the Arctic Circle, was chosen for its refrigerating properties. But Svalbard is not enough. There is more to life than seeds, and the vault itself is not invulnerable to climate change.

Harris Lewin is a biologist at the University of California, Davis. His ambition, and it is vast, is to decode the genome sequence of all eukaryotes on Earth. The digital sequence of a species cannot be destroyed by climate change, and it can be shared around the world. So far we've sequenced 2,500 species. Lewin's Earth BioGenome Project is open access. It will recognize the rights of countries to the genetic information their ecosystems contain, following an agreement known as the Nagoya Protocol. Lewin estimates the total cost at $4.7 billion. When you compare that to the cost of sequencing the human genome ($4.8 billion in today's money), you become aware just how fast genetics technology has come down in price. And the amazing bargain we could get.

With our windfall we can not only fund Lewin's BioGenome Project, but expand it to cover all life-forms. With the rate of extinction so high, it's hard to think of a more vital endeavor. The Human Genome Project promises to revolutionize medicine in a wave of change that is still only just gathering pace. Alongside that, the fields of forensics, archaeology, and bioinformatics have changed beyond recognition. And every $1 of public money invested in the Human Genome Project generated more than $140 in economic activity.

The Earth BioGenome Project could bring unimaginable advances in science—improved medicines and materials, biofuels, and crops. Having a vast genetic databank of millions more species will improve our understanding of the evolution of life on Earth. We will start to understand why some species are more vulnerable to extinction, and which we need to focus on first. Lewin says that, just as we couldn't predict what would happen as we unraveled the human genome, neither can we see where the BioGenome Project will take us. Advances in renewable energy, agriculture, engineering, artificial intelligence. . . .

Alongside this vital project, we will invest in understanding methods of gene editing that can be applied to nonhuman species in order to safeguard them for the future, and how we can genetically augment existing ecosystems to increase their resilience, while at the same time preserving their genetic diversity. For example, we might want to boost species tolerance to salt water, as sea levels rise and encroach terrestrial ecosystems. Certainly we'll want to do this for crops, and find ways to make them flood- and drought-tolerant, and we'll look at improving our agricultural systems in chapter 8. We'll have to do this with corals, too, if we are to have any hope of keeping them. With the number of possible applications and the scale of the basic project, I think Lewin's $4.7 billion is probably an underestimate, so we'll invest $10 billion to make sure.

❧

THE GENOME PROJECT will help identify which species are most vulnerable. But there is a more important and pressing way to do this. Stuart Butchart is an ecologist at Bird Life International in Cambridge, England. In 2012, he published a paper, with a team of international scientists, that crunched a mass of data from conservation biology and calculated how much it would cost to safeguard globally threatened species. All of them.

Since birds are the best cataloged of all species, Butchart and colleagues used data from birds to determine the cost of preventing the extinction of threatened species and of meeting the targets agreed: Having 17 percent of land and 10 percent of marine areas properly protected.

By estimating the cost of protecting the birds' habitats and the minimum it would take to down-list threatened species to the next lower level on the International Union for Conservation of Nature (IUCN) Red List—say, from critically endangered to endangered—the team came out with a figure of between $875 million and $1.23 billion a year. They were able to broaden this estimate to other threatened species—mammals, amphibians, and a selection of fish, plant, and invertebrate groups. Managing and protecting land for all these taxonomic groups, they estimated, would cost $76.1 billion per year.[31] For birds, roughly half the cost goes on establishing the protected area—basically buying the land. We've seen how much is being lost in terms of ecosystem services. This is money well spent.

Seventy-six billion is an order of magnitude more than what is currently spent, and we could fund it quite easily. But what it doesn't take into account is providing livelihoods for people in those protected areas. Adding this may well inflate the cost of creating the protected zones. The core spend would go toward identifying and expanding areas of land and ocean that are particular strongholds of species richness, and on managing and protecting those sites. These are referred to as the Key Biodiversity Areas (KBAs).[32] We know where some of them are, but we urgently need to complete the identification of these areas, globally, and for governments to sign up to protect them. Our money should help persuade them. Where KBAs that aren't currently protected are identified, we need to invest in protecting them, while working with local and indigenous people to build their ability to engage with conservation and create community-managed nature reserves that benefit them, through REDD+ and other schemes.

There is no time to lose here. It's not even that ecosystems are fragmenting and disappearing, or being converted to agriculture or housing; it's that it's harder to get them back once they are gone. Politically and socially it's harder, not just ecologically. Once land is unwilded, it's very hard to rewild. As Brendan Costelloe of the British Ecological Society points out, the history of Europe shows this, and so we have to act fast, especially in Africa, where much of the wildlife requires large territories and is prone to conflict with humans. Again, our intervention must also promote the social benefit of the people on that land. We have to consider the needs for food production (we'll come back to this in chapter 7) and how poorer countries can improve living standards. Currently the best option to make a living for many people in low-income countries is to exploit natural resources. Of course, much as low- and middle-income economies might resist signing up to carbon emissions reductions by arguing that Western countries got rich by polluting the atmosphere, so why should they pay the price, a similar argument might be made that European countries destroyed their own natural heritage—and through their colonies, that of many other countries—but that's a bit of a cut-off-your-nose-to-spite-your-face argument.

＊

THERE'S NO GETTING AWAY from being the rich, well-meaning white do-gooder here—that's literally what I am—but I can at least try to recognize that and assign the money to local representatives. The whole planet needs some attention. The Biodiversity Intactness Index measures the integrity of the world's land, everything from temperate forests in North America to the vast plains of Russia. Some 58 percent has "severely compromised" biodiversity. The world's oceans are similarly impacted, with just 13 percent classed as "wild." In other words, most of the blue of our planet, including almost all of the Atlantic and northern Pacific, has been disrupted, degraded, and stressed by human activity. Currently less than 5 percent of marine wilderness is protected.

There's a classic example of how diverse ecosystems are stronger ecosystems. Yellowstone National Park is a vast wilderness park in (mostly) Wyoming, covering nearly 3,500 square miles. Established in 1872 as the first of the national parks in the US, wolves were hunted out of the entire area in the early twentieth century, mostly over safety concerns. In 1995, they were reintroduced after ecologists argued that top predators such as wolves have a cascading influence over the rest of the ecosystem, and that without their presence the cascade fails to take place.[33] In Yellowstone, the extermination of the wolf had changed the entire landscape. The elk population, unhindered by wolfish predation, had grown, devouring vast amounts of seedlings and saplings, edging out bison and overgrazing young trees. Beavers, without the trees to build their dams, disappeared. Yellowstone without wolves had become more of a grassy wilderness, and a quieter place.

When a wolf pack was reintroduced to the park, the number of elk went down, which allowed more trees to grow—willow, aspen, and cottonwood. Beavers, reintroduced at around the same time, started engineering the landscape, and the regulation of water flow raised the water table and created new habitats for fish. New insect species returned with the trees, and bird numbers increased with the new nesting and foraging opportunities. Coyote numbers dropped and pronghorn antelope numbers grew. Such is the story told in textbooks. Biodiversity increased, though the response of the ecosystem was not magic or automatic, and needed careful human management. But the story *is* compelling. It just needs a large area to work. It's feasible in a big national park, but what about where the human population density is higher?

The Sunderbans of Bangladesh and India is an epic and magnificent mangrove ecosystem formed as the Ganges and Brahmaputra rivers fan out into the Indian Ocean. The top predator is the tiger. It keeps the numbers of spotted deer, macaques, and boar at a level that promotes a greater diversity of plant and animal life in the mangrove. The Sunderbans covers some 4,000 square miles, a bit more than Yellowstone, but unlike the American park, where no one lives, some four million people live in the Sunderbans. Some parts are supposedly reserved for the tigers, but people have needs, too, and that leads to inter-species conflict.

When I visited, everyone I met had a tiger story. I spoke with relatives of a woman killed a few weeks before by a tiger as she was out collecting crabs in an area fenced off for tigers. Her blood-stained sari had been found in the mud, surrounded by tiger pugmarks (footprints). The body was never found, but her sari and bangles were brought back to the village and cremated. "We don't feel enmity with the tigers," one of the villagers told me. "But for our livelihood we have to go there." Another villager showed me the wooden fence she'd seen a tiger leap over, taking the woman's neighbor.

The late biologist E. O. Wilson said we needed to allocate half of the entire land area of the planet to conservation.[34] The idea was adopted by an international group of scientists who propose that we move to protect a third of all ecosystems by 2030 and half by 2050.[35] But my encounter in the Sunderbans shows the problem that proponents of half-Earth conservation face: When we identify regions that need protecting, we need to also come up with a plan for what to do with the people who live there. I didn't see a tiger while I was there, only a fresh pugmark in the mud. But in the river I saw an Irrawaddy dolphin. They have a stubby dorsal fin and a rounded, snoutless head; they look like baby whales.

In other places where top predators have disappeared, it's hardly conceivable to reintroduce them. In West Sussex in southern England—not far from London's Gatwick Airport—there is a 3,500-acre estate owned by Isabella Tree and Charlie Burrell. In 2001, after failing to eke out a decent amount of traditional crops from the stodgy clay, they started using animals such as wild pigs, cattle, and deer to revert the farmed landscape back to a semi-wild state, and now animals roam the area, and the biodiversity has leaped up.

Knepp is a large estate when compared to the size of my back garden, but 3,500 acres is 5 square miles. A fraction of Yellowstone or the Sunderbans. It is not big enough to support a wolf pack, or bears; the introduction of top predators is not an option.

Humans take that role, culling a number of animals each year. The estate, transformed from claggy farmland, now supports a range of rare species such as turtle doves, nightingales, peregrine falcons, and purple emperor butterflies. In 2020, wild storks nested and reared young at Knepp, the first time the birds had bred in Britain for hundreds of years.[36]

What we need to do is threefold.

First, we must identify the most important areas for wildlife conservation. Ecologists call these Key Biodiversity Areas (KBAs), and they are the top priority when it comes to protection. Some of the hard work has been done for us: Eric Dinerstein of the environmental organization RESOLVE, and colleagues, have mapped protected areas of land, areas that are unprotected but that contain rare species, clusters of species and large mammals, regions that are relatively untouched by human activity, and areas that are particular important carbon stores. They came up with a map of land that needs protecting, a patchwork they call the Global Safety Net, which coincidentally to E. O. Wilson's suggestion, covers 50 percent of all land on the planet.[37]

Second, we must find ways to create conservation-protected zones that allow indigenous people to live freely in these areas. And, third, we must do what we can to promote biodiversity in smaller areas, as has been achieved on the Knepp estate.

Governments from seventy-two countries signed a Leaders' Pledge for Nature in September 2020, promising to address environmental destruction, and "reverse" biodiversity loss by 2030.[38] Pledges are very welcome—we like pledges. But our money will help governments keep their word.

⟁

HUMANS ARE A MAJOR FORCE for homogenization—a phenomenon also known as "McDonald's-ification" or the "Starbucks

Effect." Essentially we make everything start to look the same. In South America, some fifty-seven genera of megafauna went extinct over a drawn-out period at the end of the last ice age. Megafauna refers to animals over ninety-seven pounds in weight—mostly large mammals, but also some giant birds. They include animals such as the giant ground sloth and the massive, elephant-like *Cuvieronius*. The grazing, seed-dispersing, and browsing function they performed was taken over by just three species: people, horses, and cows. In North America, ninety genera went extinct at the hands of humans, including saber-toothed tigers, dire wolves, giant tortoises, giant condors, and the evocatively named terror bird (an eight-foot-tall, 330-pound, ferociously beaked flightless bird).[39] Again, humans and the same two domestic species took over the lost functions. A similar story played out in Australia and Europe.

And the story continues today. Apologies for the populist overtones, but people are facilitating the "invasion" of ecosystems around the world by "foreign" species. Cats, rats, and goats are the worst offenders, but insects (such as the terribly destructive oak processionary moth), parasites (such as the one that causes avian malaria), microorganisms (such as the chytrid fungus that is devastating amphibian populations), and ornamental garden plants (such as the *Alicenia miconia* tree that has devastated the native plants of Tahiti) do huge damage. One of the conclusions of the Millennium Ecosystem Assessment, an in-depth analysis of the state of the world's habitats, was that invasive species are a major threat to ecosystem integrity. The major route for invasion used to be human trade, and climate change is exacerbating this by opening up entirely new ecosystems: The Arctic and Antarctic, for example, are being colonized by non-native species as shipping routes change and warmer-water species gain a foothold. One paper found that invasive species formed the second-biggest threat to Red List species, after direct threats such as hunting and harvesting.[40]

A report by the Intergovernmental Science-Policy Platform on Biodiversity and Ecosystem Services (IPBES) showed that direct destruction of ecosystems was at least as big a threat to the biosphere as climate change. People don't seem to have taken this on board like they are starting to consider climate change. We know that it's bad to clear rainforests, and we are roused by images of burned koalas, or starving polar bears, or an orangutan stranded in a solitary remaining rainforest tree. But the impact of extinction on ecosystem services—the connection between decreasing diversity and a weakened biosphere—is not always appreciated. Many countries signed up to the twenty Aichi biodiversity targets, set by the UN's Strategic Plan for Biodiversity 2011–2020, but the targets were completely fluffed. Six were partially achieved, while fourteen were utterly missed.[41] Life on Earth is in "pervasive decline," said a report. "Only immediate transformation of global business-as-usual economies and operations will sustain nature as we know it, and us, into the future," wrote the authors, in the journal *Science*. "Every delay will make the task even harder."[42]

In October 2021, in a meeting in Kunming, China, delayed several times because of coronavirus, new targets were set. For all our sakes, and for the future of biodiversity on the planet, the targets must be ambitious, and must be embraced by powerful countries. The coronavirus crisis, as economically devastating as it has been, cannot be used as an excuse for inaction. Can we hope that China steps up and provides the global ecological leadership the planet needs? China is hardly leading the way with a transition to renewable energy, with huge amounts of fossil fuel power stations still coming online. Still, we can hope. There are promising signs, as we saw with Chinese government commitments to peaking carbon emissions around 2030. Three quarters of the land surface and two thirds of the ocean have been impacted by our activity, remember. A lucrative injection of funding here may help push China in the right direction.

Astronomers sometimes talk about "the Great Filter" as a way of explaining why we don't see alien civilizations in our galaxy. The great filter is an event—usually thought of as an asteroid strike, or climate change, or nuclear war—that destroys an otherwise intelligent species. But it might just as well be the destruction of planetary life support through the loss of biodiversity.

Achieved

Transformative change in land and sea use, the emergence of a sustainable form of global development, the restoration of planetary life support and the preservation of diverse life on Earth; the avoidance of the collapse of human civilization.

Money spent

Support for Kunming 2021 biodiversity targets:	$100 billion
Creation of Canopy, global forestry protection agency:	$100 billion
Projects to "wild" abandoned and degraded land:	$50 billion
Verification, purchase, and protection of key biodiversity areas for five years:	$500 billion
Support for coastal communities near coral and research into climate-smart corals/coral transplantation methods:	$100 billion
Catalog and identify all species:	$10 billion
Total:	**$860 billion**

Charles M. Duke Jr. on the Apollo 16
mission, collecting lunar samples at
Plum Crater, April 1972.

Settle Off-Planet

AIM: To establish a lasting human settlement on the Moon. To open up the rest of the solar system to greater exploration, to eventually relieve ecological pressure on Earth, and to begin the next stage of the human journey. We will create an international organization, the Terran Alliance, aimed at building an inclusive community on the lunar surface that adheres to the sentiment of the Moon Treaty of 1979: The Moon is for everyone, not just billionaires and mining companies.

WHEN I WAS A CHILD, I had a poster in my bedroom of a Space Shuttle piggybacking on a Boeing 747. I liked it because it was a spaceship, and spaceships are inherently cool. They are about the future. Unfortunately, that was going on forty years ago, and human spaceflight of the kind I dreamed of remains the future. This chapter is about changing that forever.

If you never got the space craze, or if you quite reasonably think that sending people to space is far too expensive when we can do everything we need with robots instead and have enough problems to worry about on Earth, I hear you. But I want to try and bring you along. This is not about going to space because I'm still pitifully clinging to childhood dreams of being an astronaut (or it's not just about that). It's about making a commitment to international cooperation, to the creation of knowledge, and to the future both of humanity and of Earth. It's to explore. Eventually, it may relieve some ecological pressure from Earth. But ultimately, it's to get to the future quicker.

It's also very doable on our budget. With the Apollo program in the 1960s, NASA developed unprecedented rocket technology, the Lunar Lander module, and other things such as life support systems and a lunar vehicle, for "only" $220 billion (in 2019 prices). Perhaps it's a symptom of my thinking so hard about $1 trillion that $220 billion seems a small amount. Apollo, one of the most extraordinary achievements of human history (so far—we're only getting started), managed a huge amount on that budget, with some 400,000 people involved in the effort, but it was only a means to an end. It was a race to get people on the Moon and, when the United States won, the wind fell from its sails. Our space program will be for the long term; with our money we can truly start the next chapter of the human story. We live on a beautiful and precious rock, but we should have gotten off the damn thing by now.

∽

My first thought was to go to Mars. After all, no one's ever been there before. It's a bold and sensational goal, one that would inspire the public: the mission to walk, for the first time, on another planet. In Elon Musk's SpaceX rhetoric, we would become a multi-planet species. Mars is almost habitable. It has an atmosphere—a thin one, granted—but it still has one. There is carbon dioxide we can use to make methane fuel, and lots of water in the form of ice that can be consumed and broken up into hydrogen and oxygen. The biological science of Mars is fascinating, too: It's quite likely that there was life on the planet in the distant past, and there may even be microbial life there now. And, to embrace a favorite of science fiction, it's possible that we could terraform the planet: Turn the cold, barren, and mostly dry Mars into a wet, warm, and habitable version of Earth.

This is a vision that is explicitly presented in Musk's SpaceX imagery and promotional materials. We have the chance to begin again, to be our better selves. Neil Armstrong, it is said, is the only person from the twentieth century whose name will be remembered in the thirtieth. Whoever first sets foot on Mars will never be forgotten, and will literally pave the way for the future of humanity.

With this kind of buildup, it's hard not to get carried away. But when I started looking into it a bit more, I reasoned myself down. It just doesn't make sense to go to Mars without going back to the Moon first. We don't want to fritter our money away, we want to use it to create something long-lasting. A sustainable base on the Moon is the way to do that, not the one-off spectacular that is all our money would get us if we went to Mars. Pascal Lee of the Mars Institute, a nonprofit research group partly funded by NASA, has calculated that a human mission to Mars could cost up to $1 trillion over twenty-five years. It's not just that getting to Mars is incredibly difficult and expensive, it's that surviving once you're there throws up massive new challenges.

NASA has thought a lot about this. It calls the unknown factors surrounding a Mars mission—the things we need to figure out before we could attempt a human mission, let alone establish a long-term settlement there—SKGs (Strategic Knowledge Gaps). They include, but aren't limited to: data about the atmospheric dynamics and weather of the planet, which is needed to enter orbit and land; information about potential past and present life on the planet, including the distribution of past and present water; and data about the back- and forward-contamination risk— that is, the biohazard posed by passing Martian material to Earth, and vice versa. We need to develop technology that will sustain human life during the twelve-month trip to and from Mars, and sustain humans on the surface for multiple Martian years. We also need information about the toxicity of the dust, and the radiation exposure risk on the surface, and to identify a safe and suitable landing site for a human mission. We don't know anything about the psychological impact of being on another planet.

To close the strategic knowledge gaps, NASA has identified dozens of GFAs (Gap Filling Activities). So couldn't we just parachute money in to fund these activities? Sure, we *could*. We could try to spend our way through the unprecedented academic, social, engineering, and scientific challenges, the bureaucracy, the unbelievable effort and attention to detail required, not to mention the ethical and geopolitical turmoil a settlement program would kick up. But the Moon has more to offer.

∾

AL WORDEN WAS AN OLD-SCHOOL *Right Stuff* astronaut. A back-up pilot to Apollo 12, he flew the 1971 Apollo 15 mission to the Moon, and was the first person ever to spacewalk. As he orbited the Moon, alone in the Apollo 15 command module, he got to a distance of 2,235 miles from his crew mates on the lunar surface, earning him a place in the *Guinness Book of Records* as the most isolated human in history. Like most of the Apollo astronauts, he was a former test pilot and the loneliness didn't seem to bother him; he said he enjoyed being in the spacecraft all by himself. But he did reveal something of a reflective nature, noting in a poem,

"Now I know why I'm here. Not for a closer look at the Moon, but to look back at our home, the Earth."

Al died in 2020. But he once told me it was essential that we go to the Moon as a staging post to Mars, if nothing else. He told me while smoking a cigarette—defiantly the nothing-can-harm-me hero archetype, the sort of person who speaks at you rather than with you, which probably comes from a lifetime of knowing that his stories are always going to be the best of anyone else in the room. But I was listening. For Al, the Moon came first.

His old friend Buzz Aldrin says the same thing. But don't jump to the conclusion that it's just because they've been to the Moon that they advocate going back. I heard similar arguments from several NASA scientists I spoke with. Mars and the Moon are almost equally as deadly to humans at the surface, as both have no magnetic field, barely have an atmosphere, and require a spacesuit. However, with the Moon you can get back to Earth relatively quickly, in only three days, if something were to go wrong. It would take anywhere between 150 to 300 days to get back to Earth from Mars. The time delay from Earth when speaking to someone on the Moon is 1.25 seconds; on Mars, it's anywhere between 4 and 24 minutes, depending on the position of the Martian orbit relative to Earth. We need to learn how to live on the Moon before we make the step to multi-planet status.

If we choose the Moon, we have the added benefit that there is currently a lot of international activity directed there, both human and robotic missions. It's been called the new Space Race. NASA's Artemis program, at an estimated $28 billion, aims to put a man, and the first woman, on the Moon by the mid-2020s. We can piggyback on that, and make our trillion go further. The Moon also opens up the rest of the solar system. The most expensive, difficult, and limiting part of space travel is the effort it takes to break free of Earth's orbit and fling yourself at your target. Rocket engineers refer to necessary change in velocity as

Delta-V, and the value is much less from the Moon than it is from Earth. It's why the Apollo program needed the biggest rocket ever built, the Saturn V, to get to the Moon from Earth, but only a tiny rocket was needed to get off the Moon and back to Earth.

Any crewed mission to Mars will have to be trailed by multiple scouting and setup missions, and any sustained buildup of resources on Mars will require hundreds of missions. It may be far more feasible, far cheaper, to build as much as what we need on the Moon and take it to Mars from there, rather than all the way from Earth direct. The Moon has large amounts of frozen water, which we could mine and use to make rocket fuel to refuel our ships, making return trips cheaper and onward missions easier. The rockets themselves won't need to be as big, so they'll be cheaper, and we'll be able to more easily send supply ships and one-off exploratory probes to locations around the solar system. So let's do it. Let's make the Moon the eighth continent.

∽

WE NEED TO ROW BACK A BIT and examine the reasons we should invest in a program to build an off-planet settlement in the first place. There are difficult ethical questions to consider, too. We need to robustly defend our choices.

I'm going to try and keep Marvin Gaye in mind here. (It's not a bad rule of thumb generally: When my wife and I got married, we played "You're All I Need to Get By" at the ceremony.) On "Inner City Blues," Marvin sang "Rockets, moon shots; spend it on the have-nots," by way of commentary on the money given to the Apollo program, while black people's problems were basically ignored. The words of that song are just as relevant today. Gil Scott-Heron said something similar in 1970: "I can't pay my doctor bills but whitey's on the Moon." While many people were enthralled by the Apollo program, many were concerned that it was a huge indulgence to look to space when there were enough problems on the ground. Nothing much has changed.

In 2018, Musk's SpaceX company launched its Falcon Heavy rocket for the first time. This is an impressive rocket, the most powerful built since the massive Saturn-Vs that took "whitey" to the Moon. Elon Musk says it has cost around $500 million to develop—bear that in mind. Of all rockets currently operational, the Falcon Heavy has the biggest payload. Musk had that first flight carry a payload containing a red Tesla roadster with the roof down, a "Starman" mannequin behind the driving wheel and the David Bowie song playing on the car stereo. The mannequin was positioned with one hand on the wheel, the other arm resting on the door. Some people were angry with what they saw as macho and patriarchal imagery and the same old middle-aged, rich, white, male-dominated agenda. The CEO of SpaceX is a woman, Gwynne Shotwell, but it didn't help that the crowds cheering in the SpaceX launch control center were almost entirely white men.

Nor does it help that Musk likes to talk about establishing colonies on Mars, and "conquering" the Moon. The language is inflammatory to some because it recalls the evils of imperial colonization and slavery. He also suggested we nuke Mars—presumably at the poles, to melt the ice there and warm the planet, in order to kick-start the terraforming process. It's unclear if he's behaving like an amped-up Bond villain to irritate his critics, or if he really wants to nuke Mars.* Either way, we can do it differently. As it happens, a group of entrepreneurs and former NASA scientists

---

* I thought briefly about trying to restart the protective magnetic field on Mars. Earth has a magnetic field generated by its hot core of liquid iron. Mars has a molten iron outer core but the inner core seems to be cold and solid. To restart the magnetic field would require melting this inner core. You'd have to drill thousands of miles into the planet and then maybe detonate nuclear warheads to melt the core. It's not really feasible any time soon.

calling themselves the Open Lunar Foundation have similar ideas—we will end up collaborating with them.[1]

Some reasons put forward for space travel, not mutually exclusive, are: to do science, to explore, to save the Earth's environment, to start independent human settlements, as insurance against catastrophe, to make fabulous amounts of money, to go for glory, to go for greed. Some of these are pie in the sky, or delusional, or deliberately misleading. I can't see how anyone is going to make money from investing in a Moon base any time soon. SpaceX is well paid from NASA delivery contracts to the International Space Station, and it is looking to make money from orphans of Apollo and other wealthy space tourists. Enthusiasts dream of prospecting for valuable minerals on the lunar surface, and of a business that could be set up refining lunar ice to make rocket fuel. But there is no space-based economy and there won't be one for decades.

Then there is the insurance idea. That's the argument that we need to start an off-planet human settlement in order to ensure the human species survives in the event of catastrophe. An asteroid strike ended the reign of the dinosaurs 66 million years ago; a super-volcanic eruption could cause an extinction event of the same scale. But, statistically, neither is likely to happen soon, despite the claims of Stephen Hawking in his final book that an asteroid impact is the greatest threat to life on Earth. To argue that we should spend money on space missions for this reason, to create a backup for the human species, seems disingenuous. Biodiversity on Earth is *genuinely* facing multiple threats, right now, as we saw in the previous chapter, and we should face them here, and fight them here, not run off to some other planetary body leaving the vast majority of people to suffer. We should acknowledge that and not try to dress up our Moon base plans as something they're not.

It is, of course, highly laudable to claim that a Moon base will save Earth's environment, which is what Jeff Bezos often says.

The founder of Amazon, and the spaceflight company Blue Origin, Bezos talks of relieving environmental pressure on Earth by shifting heavy industry to the Moon. Blue Origin has developed a lunar lander and is building a rocket to get it there. Bezos wants to move heavy, polluting industry off Earth and relocate it on the Moon, eventually making Earth a residential zone with a bit of light industry and far less pressure on the environment. Some without stars in their eyes see Bezos's plan as extending his power and wealth ever further, to an end point where Blue Origin produces goods on the Moon that Amazon delivers on Earth.[2]

Again, whatever the outcomes of Bezos's grand plan, they are far in the future. To some people, our trashing of the Earth environment seriously compromises our right to start settling on another celestial body; it is hardly acceptable to start strip-mining the Moon when our home planet is in such a state. We'll come to this—I take it very seriously—but for now the point is that Bezos's goals are long-term; we want something more immediate to focus on. I need a simple answer for when people ask why I'm spending so much money on the Moon.

CHINA ACHIEVED A WORLD FIRST in 2019, when it successfully landed a rover on the far side of the Moon. Since the far side of the Moon is permanently facing away from Earth, it sits in a cone of radio silence, and any spacecraft entering the cone is cut off. It's why many scientists dream of setting up experiments on the far side, where they are shielded from radio interference from Earth. No one knows of another place within many light-years that is so protected. So China's Chang'e 4 probe had to navigate autonomously, without human control, once it passed to the far side. This was a hugely impressive achievement for a robot, and it's missions such as this that make some people think all our exploration can be carried out by robots. China is a major player, but India, Israel, Japan, and the European Space Agency (ESA) have their own lunar missions underway or in preparation.

In one sense all this activity is good. There is finally a momentum again behind space exploration. But it's feeling too fragmented, too vulnerable to whim and market. The US Constellation program was set up under the George W. Bush administration to put people on the Moon, but it was canceled by Barack Obama. Obama promised $6 billion to go toward crewed missions to Mars in the 2030s, but that was canceled by Donald Trump. In 2019, Vice President Mike Pence announced that the US would return to the Moon by 2024, but then Trump apparently switched his affections back to Mars. NASA is nevertheless working on the Artemis program. The agency's budget is around $20 billion a year, but it will need at least another $8 billion a year to achieve another Moon landing. We can smooth over the short termism and the vagaries of the presidential cycle that hinders NASA. We can also use our clout to ensure that space does not become dominated by billionaires.

The space race of the 1950s and 1960s was for glory, prestige, and power. When the Soviet Union launched the first satellite, Sputnik 1, into orbit in 1957, then-Democrat majority leader Lyndon Johnson stoked the idea that it was an affront to American prestige. "Control of space means control of the world," he said. In 1961, after Yuri Gagarin had become the first person to orbit the planet, President John F. Kennedy quickly got behind the space endeavor. He wanted something more dramatic than the Russians, and committed the US to landing a man on the Moon within ten years, famously invoking the human thirst for exploration and new achievements: "We choose to go to the Moon in this decade and do the other things, not because they are easy, but because they are hard." Damn right it was hard. Gus Grissom, the commander of Apollo 1 (he died tragically in a fire on the launch pad), put it like this: "It's as if somebody had said, 'Let's build New York City overnight.'"

Apollo achieved greatness, but it didn't give the United States "control" of space. And glory alone isn't enough to sustain something long-term. The Moon Treaty of 1978 proclaimed that the Moon is "the common heritage of all mankind," but neither the US nor the Soviet Union signed it, because they were concerned it would rob them of future control, glory, and economic return.

⌖

THERE IS SCIENCE. We've heard how the far side is unpolluted by Earth's radio noise. The Moon itself is also a time capsule. It is geologically inert, with no tectonic activity churning its rocks as on Earth, meaning that it is a pristine record of how things were in the early days of our solar system. And there is applied science. When we are there, we will be forced to learn things that we should have learned years ago on Earth: how to use and recycle things efficiently, and how to use renewable energy for all our needs. This know-how will be useful back on Earth.

Getting to the Moon is currently a competition, another race. But, once we are there, once there are settlements on the Moon, the lunar people will surely want to cooperate together. It will be in their best interest. We could shape that cooperation now. Imagine if we spent hundreds of billions of dollars creating an alliance between private space firms and nation states. We could combine all our expertise and build toward a goal where all countries had a stake. SpaceX is developing Starship, the successor rocket to the Falcon Heavy, which is due to fly around the Moon by 2023 and which has been referred to in earlier incarnations as the Mars Colonial Transporter. China's heavy-launch system of rockets is called Long March; Blue Origin has the New Glenn range; and NASA itself has the much-delayed Space

Launch System. If we form an umbrella organization, the Terran Alliance, we can foster a spirit of togetherness and knowledge sharing, while maintaining a degree of independence, allowing individual nation-states and private companies a certain autonomy, and agreeing on a new Moon Treaty for exploration and exploitation of lunar resources.

We can work out the details later and the amount of funding each member of the TA receives, but something like $50 billion each for SpaceX and Blue Origin, and $100 billion each to NASA and China, seems like a good starting point. We should also invest $50 billion in ESA, and, importantly, $50 billion in the African Space Agency. The African Union established a space agency in 2017, and in 2019 voted to base its headquarters in Egypt. Kenya launched a satellite in 2018, and Ethiopia in 2019, building on previous collaboration with Europe. With their greater vulnerability to climate change, it is vital that African countries have access to satellite data, which will help with deforestation protection and agricultural planning. Our investment will guarantee this development and recruitment for the Moon program from across the continent. This time it won't be only "whitey" on the Moon.

In the mid-term, a settlement on the Moon will help with our exploration of the rest of the solar system, and in the further future we can look to the Bezos dream of industry and power generation on the Moon.

As we've seen in chapter 3, we need to transition as soon as possible from a fossil-fuel-based society and economy to one based on renewable energy. Some reports suggest that to transition to 100 percent renewable energy will be difficult in terms of the material resources required. Specifically, the minerals and lithium needed to make enough batteries will require huge amounts of mining activity. It will also require land that we can ill afford to spare. If we plan to get the materials we need from space, from asteroids and the Moon, so the argument goes, we can start to restore Earth. To my mind it's better to dig holes in the Moon and consume asteroids than further erode our own planet. But proposals to industrialize the space environment leave many people aghast. Aren't we going to repeat the same mistakes and exploitation that we've done on Earth, to terrible cost?

I would *like* to argue that there's nothing wrong with mining dead asteroids and even the Moon, but then we *thought* there was nothing wrong with plowing the rainforest and farming cattle on an industrial scale. As a society we actually still accept those destructive processes. My impulsive response to those who want to exploit the Moon is: Go for it, better there than Mars, which may well contain archaeological evidence of past life and may even harbor existing life-forms.

Once we've kickstarted the process of settlement, it should become self-sustaining. Phil Metzger and Julie Brisset of the Florida Space Institute have modeled the business case for setting up a water-mining operation on the Moon, aimed at producing rocket fuel which could be sold at space stations in low Earth orbit. Metzger and Brisset's study, which was funded by the United Launch Alliance, finds that with a little investment (they assume from NASA, but it could be from us) a mining operation could get up and running. A fledgling space-based economy could follow. They argue that a financial payback could come sooner than expected, in a matter of decades. To my mind the more exciting point is that the environmental payback could be incalculable.

Of all the environmental pressures we face on Earth, there's one that I hadn't appreciated fully until I started thinking about the potential benefits of a lunar presence. It goes without saying that computers are everywhere. It's not that we take them for granted, it's that the apparently free access we have to the internet removes the idea of energy consumption from our consideration.

We don't pay for this electricity, so we don't think about it, and the energy use mounts up.

As I write, in 2020, there are at least 10 billion entities—toasters and Fitbits and fridges and cars and robots and other automated systems—using the internet. That's in addition to the people using it. Combined, the world's ICT accounts for 2.5 percent of global emissions (1.5 billion tons of CO_2 equivalent, annually). Bitcoin and other cryptocurrencies alone burn through 75 million tons of CO_2 equivalent each year. As computing power and usage increase, so does the energy consumption and the carbon emissions. This needs to become sustainable.

One answer is to shift some of the computing infrastructure off-planet. Energy and minerals can be provided from the Moon, without further eroding Earth's biosphere. Solar power is available constantly in some locations on the Moon and, long-term, we can look to relieve the burden of energy production on Earth by delivering some of it from space. Space-based solar power—where giant arrays of solar panels capture sunlight and beam the electricity down to Earth in the form of focused microwaves—has been written about in science fiction for decades, but at Caltech Harry Atwater's team is working toward this. It might not work, but even if it doesn't we should be able to shift a lot of computer-processing hardware to the Moon. It's not the most romantic of images, but the Moon might end up a massive server farm for Earth.

Is it enough to say that we should go to space because we are explorers, and exploring adds to the stock of human knowledge? I don't know if that is enough, but it's inevitable that humans will go wherever we can. The fact that we can go to space feels almost miraculous, but, since we *can* do it, we will. Some people see this aspect of human nature and say we're like an infection, forever reproducing and spreading and harming the places we go to. But all life evolved to do this; it's not unique to humans. What is unique to us is that we create knowledge, and we can learn to do things better. What we can do with our money is ensure that we go to space for the right reasons, and try not to behave like an infection. If it is left to market forces, or to national pride, or to the whim of billionaires, space travel becomes something different.

Now that's settled, we need to get into some details. First, deciding where on the Moon we go.

∽

I HAVE A SMALL TELESCOPE at home in London. What with the light pollution and the low power of the scope, pretty much the only thing worth looking at is the Moon. But what a sight. I never get tired of staring at the craters and the impact rays, the streaks of debris thrown out by meteor impacts, the jagged light-colored anorthosite highlands, and the dark basalt lava fields of the maria, the "oceans" that make up the Man in the Moon.

The best time for Moon watching is when it is waxing or waning, as then it's not too bright and the peaks and craters picked out by the sunlight on the terminator—the boundary between dark and light—bring the Moon into startling three-dimensional relief and show off the surprising beauty of the terrain. I bought myself a large map of the Moon so I can put names to features; it's had the effect of ever so slightly inducing an affinity with the millions of people throughout history who have gazed up in wonder.

Easily seen on the Moon's western face is the Aristarchus crater, a twenty-five-mile-diameter impact crater deeper than the Grand Canyon which is the brightest formation on the lunar surface. Visible to the naked eye, the feature is bright because it is only 450 million years old, young for the Moon, which means there has been little time for imperceptibly slow space weathering to darken it. It is named after the ancient Greek astronomer Aristarchus of Samos, the first to argue that the Earth orbited the

Sun, and not the other way around. Another good place to find is the delicately colored Palus Somni, the Marsh of Sleep, on the eastern edge of the Sea of Tranquillity. Sink into it before bed and absorb its somnolent power.

What I can't see is one of the most promising places where we might want to build a human settlement. Shackleton crater at the South Pole is an impact crater 13 miles in diameter, 2.6 miles deep, and some 3.6 billion years old. Viewed from Earth it appears "under" the Moon, so we can only see its rim, but NASA and others have identified it as a prime candidate for a lunar base. Because of its orientation with respect to the Sun, the peaks of the crater rim are bathed in almost continuous sunlight, while by contrast the interior has never seen sunlight, and is perpetually dark and cold, a crater of eternal darkness. Solar panels on the "peaks of eternal light" would provide continuous electricity, and ice that may have built up in the crater could be mined for drinking water and to make rocket fuel and breathable air.

A source of water is key. Water is heavy to transport from Earth, and on the Moon is the most valuable of commodities. Of all the adventures and experiences and insights that astronaut Scott Kelly had spending his record-breaking year on the International Space Station, to me the most memorable comes on his return home to Houston. Kelly walked straight through his house, into his garden and into the swimming pool, still in his flight suit. "The sensation of being immersed in water for the first time in a year is impossible to describe," he said. "I'll never take water for granted again."[3]

ESA has been working with the Massachusetts Institute of Technology and specialist architect firm SOM to design a settlement at Shackleton, which it calls the Moon Village. (It makes me think about how we should name our settlement. The crater is named after Ernest Shackleton, the British Antarctic explorer, but we might want to have a name that doesn't hark back to British imperial power.)

There are several other promising sites, which we will scout in detail with cheap satellite and robot missions before making our final decision. In the vast region of basalt known as the Oceanus Procellarum—just south, in fact, of the Aristarchus crater—are the Marius Hills. Scientists had known for some time that there was what they called a "skylight" in this region, an apparent opening into an underground cavity. Then, in 2016, observations made by the Japanese lunar orbiter Kaguya and NASA's GRAIL (Gravity Recovery and Interior Laboratory) mission indicated that the skylight opened into a vast cave.

Caves on Earth are typically formed by water, but on the Moon they are formed by lava flow. When the Moon was volcanically active and lava drained across its surface, sometimes the top cooled enough to form a crust, while still flowing underneath. The crust forms a hard roof and, if the lava drains out of the channel, an empty tube is left behind. The talk of them as tubes puts me in mind of the London Underground, but the tube at Marius Hills could be large enough to contain Philadelphia, said researchers at Purdue University, Indiana. Less of a tube, more of a gigantic cavern. Imagine the heating costs. But lava tubes could be a good long-term solution to some serious problems astronauts will encounter on the lunar surface. Since the Moon has no atmosphere or magnetic field, it is completely exposed to cosmic rays and solar radiation. For the duration of an Apollo-style walk on the surface, it's tolerable, but for longer durations, and certainly for spending months and years on the Moon, astronauts will need substantial shielding.

In ESA's Moon Village proposal, the inflatable modular apartments are protected with "regolith-based protective shells" to defend against extreme temperatures, meteorites, and radiation. Effectively, the blow-up homes have Moon soil—regolith—piled up over them. (You can imagine a robot dump truck doing the job.) This is a good initial solution, but in the longer run we want to move, Hobbit-like, to the tunnels. It might be advisable to build rotating homes for new arrivals, where the centrifugal force restores an Earth-like feeling of gravity, but once you're used to moving around in microgravity you can have some real fun.

In his novel *Red Moon*, Kim Stanley Robinson describes settlements on the Moon, mostly built by the Americans and the Chinese, but some, such as the Free Crater, are independently financed. Here's the scene where the protagonists first visit it:

> The entire space of the crater was aerated and heated, and brightly lit by mirrors and floodlights set all around the rim. From the platform's edge they could look down and see that the space between the dome and the crater floor was occupied by scores of hanging platforms. . . . An aerial town; and people, tiny in the distance, were jumping from one place to the next, swinging like apes or monkeys.

Of course, we don't know what microgravity will do to our musculature and bone density or how we will cope long-term, but . . . we'll be able to leap like gibbons. We will practically be able to *fly* on the Moon.

GOING TO THE MOON—more than that, settling on the Moon— will hugely increase human knowledge. We will learn to live in a sustainable way with a near-circular economy and without wasting stuff: recycling materials, using renewable energy, being efficient with water and oxygen; building safe homes, mining resources from the polar regions and eventually from asteroids; installing scientific equipment on the far side and performing experiments and making observations of the universe that are impossible from Earth. The Moon will then become a stepping-stone to the rest of the solar system, to Mars and its moons, then beyond to the asteroid belt and the moons of Jupiter and Saturn. It will eventually relieve the pressure on Earth by reducing the number of polluting industries that are based there.

Getting off the planet could save it, and at the same time open the door to the future. In 1972, astronaut Gene Cernan left the Taurus–Littrow valley on the eastern edge of the Sea of Serenity and reentered his lunar lander, and no human has since been back. Cernan would later remark on the peak of human exploration the Space Race had reached and the sudden curtailment that followed. "Apollo came before its time," he said. "President Kennedy reached far into the twenty-first century, grabbed a decade of time and slipped it neatly into the 1960s and 1970s." We can enable that vision of the twenty-first century to become real.

Starting a viable, self-sustaining lunar economy is for a bit further down the line, and will be something that we'll work out as we start the program. Our investment, primarily, is in human cooperation.

The International Space Station is currently being phased out. Some people have criticized its worth in terms of clear scientific discoveries and advances, but as a collaboration of five space agencies (from the US, Canada, Russia, Japan, and Europe), and as a long-running experiment on how to live in space, it's been a resounding success. People from eighteen countries have visited since it went into operation in November 2000. The space station is now being commercialized—NASA will start offering holidays on it ($35,000 per night, not including transport to and from low Earth orbit)—but its days are numbered. It may last until 2024, perhaps a few years longer.

Collaboration elsewhere is winding up. Since the Space Shuttle program ended, NASA has relied on Russia and its Soyuz rockets to get its astronauts to the space station. But SpaceX was awarded a contract by NASA to deliver astronauts to space, and successfully launched its Dragon capsule, with people aboard, in June 2020. The Commercial Crew Program is designed to wean NASA away from reliance on Russia, so this era of US–Russian collaboration is coming to an end.

Our investment will be a reaction away from the lurch to populism and nationalism and isolationism. It is as far from that as you can get. It is an alliance of peoples, of nations; it is literally and figuratively outward-looking. It is the opposite of slogans such as "America First." It is optimistic about the future, and it is investing not only in the hardware and the machinery needed to get to the Moon and build homes on it, but in human cooperation, in a counterweight to aggression and nationalism. The Terran Alliance will not govern, but it will develop lunar prospects in the broadest terms—tourism, art, mining, tunneling, habitat construction, science on the far side. It is about sustaining and nurturing the best in human cooperation.

Achieved

The establishment, within a decade, of a permanent and diverse civilian and scientific settlement on the Moon. We will work toward relieving ecological pressure on Earth, increasing scientific knowledge, exploring the rest of the solar system, and creating a beacon of hope and international cooperation that will be seen by everyone on Earth.

Money spent

Creation of the Terran Alliance: . . . . $400 billion

Development of rocket transport system and
Lunar Lander module: . . . . $30 billion

Construction of multiple rockets: . . . . $50 billion

Cost per return flight: $600 million; 100 launches: . . . . $60 billion

Power generation (photovoltaics): . . . . $10 billion

Robot and vehicle support on the lunar surface: . . . . $50 billion

Food and water, farming and life-support systems: . . . . . $100 billion

Architecture, tunneling, and construction costs,
including tourist facilities: . . . . $100 billion

Infrastructure (road and rail links): . . . . $10 billion

Scientific equipment, including telescopes: . . . . $50 billion

Mining and fuel production on the Moon: . . . . $50 billion

Total: . . . .**$910 billion**

E. T. the Extra-Terrestrial—still
the world's most famous alien.

Find Some Aliens

AIM: To cure our cosmic loneliness through the discovery of
extraterrestrial life. To answer fundamental questions about
what even constitutes life. To learn more about our cosmic
neighborhood through the construction of new telescopes
in space, on the Moon, and on Earth, and through robot
missions to Venus, Mars, and the interesting moons of the
outer solar system.

IF WE EVER DISCOVER EXTRATERRESTRIAL LIFE, the Vatican, at
least, will be ready. In 2018, following the discovery of transiently
liquid water on Mars, Guy Consolmagno, the Pope's chief astronomer, wrote in the Vatican newspaper: "We now know that there
is liquid water and we have evidence that may not prove there's
life but is certainly consistent with some kinds of life-forms."[1]

The Vatican is an enthusiast for aliens. In 2015, an apparently
Earth-like exoplanet was discovered in the constellation Cygnus,
1,402 light-years from Earth, and the then–chief astronomer,
José Gabriel Funes, said it was great news. "It is probable that

there was life, and perhaps a form of intelligent life," he said of the planet known as Kepler-452b, going somewhat further than any scientist in his speculation.[2] There is no conflict between the existence of alien life and Catholicism, Father Funes considers, as any aliens would still have been created by a universally powerful God.[3]

The yearning to discover if we are alone in the cosmos is incredibly powerful. Imagine being able to provide an answer. Imagine being able to tell that story. Perhaps it would do for philosophy, and for our view of our planet, what Darwin's theory of evolution did for our view of our place in nature. It would connect us directly with the rest of the cosmos. It would, one hopes, inspire a firm and meaningful commitment to careful stewardship of this and other planets. Perhaps, to save our world, we need to recognize one of two things. First, that there are other forms of life out there; in which case, we need to communicate, or at least investigate. Alternatively, we may find no evidence of life elsewhere. In which case, as the only known life in the universe, we have a cosmic duty to save ourselves.

There's no guarantee that we could find alien life, even with a trillion-dollar investment, but the odds aren't bad, and our money will go a long way. Literally. The clouds of Venus are looking very interesting right now, and astrobiologists also think there's a pretty good chance of finding evidence for past or present life on Mars. If not there, then on Jupiter's moon Europa, or Saturn's moons, Titan or Enceladus. All three moons have conditions which could be suitable for life. Probably only a basic form, granted, but can you imagine the consequences of discovering it?

We'll look at the best locations in turn, moving outward in order from the Sun: first Venus, then Mars, then Europa, and finally Titan and Enceladus. One warning: Even though these are our neighbors, it takes years to get to them, so we're looking at a fairly long-term project before we have firm answers. Hints and suggestive evidence is one thing, but to declare the existence of alien life we need very strong evidence.

And we'll also be looking even further afield. Searching for life on exoplanets—planets outside our solar system—is harder, as they are so much more distant. The closest is Proxima b, just over four light-years away, but most are further still.*[4] We will invest in plans to send both crewed and uncrewed spacecraft on interstellar missions, but here our main focus will be on the next generation of telescopes that will be able to look in detail at the rocky planets around stars in habitable zones, the region where temperatures are just right for water to exist in liquid form. These are the places most conducive to life as we know it. The atmosphere of these exoplanets can be examined for chemical signs of life. By a rough calculation, there are fifty billion planets in our galaxy that are potentially habitable.

Let's get out there. Let's bring an end, one way or another, to cosmic loneliness.

∽

VENUS, THE SECOND PLANET FROM THE SUN, shines bright and beautiful in the sky because it is so thick with reflective clouds. Its atmosphere is made up almost completely of carbon dioxide, which creates an intense greenhouse effect and hellish temperatures on the surface, reaching nearly 900°F.

We can't conceive of anything living there. But in the clouds themselves the temperatures are much milder, allowing water to exist as droplets, and it is here, in these droplets, where life might find a way. We know that bacteria live in the atmosphere of Earth. We also know that they can seed rain—they are rainmaking

---

* One light-year is 9.46 trillion kilometers, the distance light travels in a year.

microbes, leading to the quite beautiful idea that rain is atmospheric bacteria's way of getting to the ground.[4]

The idea that something similar might be happening on Venus never occurred to most astrobiologists, who are focused on Mars and the moons of Jupiter and Saturn, or if it did, they dismissed it. Not so Sara Seager, at the Massachusetts Institute of Technology, and Jane Greaves at the University of Cardiff, who a few years ago started thinking about a simple molecule called phosphine, made of phosphorus and hydrogen. Simple though it is, on Earth it is only made through biological processes, so its detection on another planet would be fascinating, possibly a biomarker, an indicator of the presence of a lifeform. When Greaves detected the gas in the clouds of Venus, it generated great excitement.[5]

It doesn't mean there's life on Venus, far from it—but it's not known what chemical or geological processes could have generated the gas, if indeed it is present. It's worth further investigation. ESA has one Venus mission, and NASA another two missions, in the planning stages. India has a Venus orbiter, Shukrayaan-1, slated to launch in 2023, and a private space company, the New Zealand's Rocket Lab, is preparing a Venus orbiter for the same year.[6] We'll give each of these projects all the funding they need.

∽

BEFORE WE GO TO MARS, the fourth planet, I want to go down Mponeng gold mine in northern South Africa, the deepest mine in the world, and one of the richest sources of gold in the world (not that we're interested in its paltry hundreds of millions of dollars of gold). From the surface it takes over an hour to reach the bottom of the mine, two and a half miles underground. In 2008, biologists discovered a species of bacterium down there, living without light or oxygen, using the radioactive decay of uranium in the rock as a source of energy, in a community made only of others of its kind.

For ecologists this was a double shock. First, because of the way the bug made its living. Other deep-sea or deep-Earth organisms usually rely on at least some chemicals ultimately derived from the Sun, but the gold-mine bacterium lived entirely cut off from the Sun or any products of photosynthesis. And, second, it lived without other species; in a community of one, it has to do everything itself.[7]

Everywhere we've looked, deep in our planet, we've found life. The latest estimate suggests that the biosphere deep below our feet is by far the biggest on the planet, with a volume twice that of all the oceans, and comprising an estimated 10^{30} microbial cells.[8] It's a staggering amount. Some bacteria, living around volcanic rifts in the sea bed, can survive at 250°F. Others found deep under the South Pacific seem to be microbes hundreds of millions of years old, which would make them easily the oldest life-forms ever discovered; perhaps they are in fact immortal.*[9] Some deep-sea organisms seem to be able to lower their energy needs to an astonishingly low level, to less than a zeptowatt of power, which is 10 to the minus 21 watts. In this way they can survive for millions of years.[10]

This sort of thing has no real bearing on whether or not there's life on other bodies in the solar system. But it gets astrobiologists fizzing with excitement, simply because finding life in extreme conditions on Earth gives them hope that there might be life off of it. The nearest we've got to actual evidence of that came in 1976, when two spacecraft landed on Mars about 3,700 miles

---

* They are as good as immortal, but it's not much of a life, as far as we can tell, proceeding with almost no change for millions of years. Kind of hellish, which is appropriate.

apart. Viking 1 and Viking 2 were NASA landers sent with an experiment expressly designed to look for alien life.

The key instruments on the landers were the labeled release (LR) life detection experiments. In the experiment, each lander collected a sample of Martian soil and added it to a nutrient broth in a test chamber. The broth contained radioactively labeled carbon, and the mixture was incubated for seven days. The air in the chamber was then tested by the instruments on board the lander. The idea was that if microbes were present in the soil and had been feeding on the nutrients in the broth, the labeled carbon would show up in the air. This *was* the case—the LR experiment reported the presence of the labeled carbon, apparently metabolized by microbial activity. But NASA and most scientists have never accepted it as proof that microbial life exists on Mars, mostly because another, complementary experiment failed to detect organic compounds. This experiment itself has been robustly criticized,[11] and the principal investigator in charge of the LR experiment, Gilbert Levin, has long been convinced that Viking did indeed discover signs of life.[12] But at the moment we don't have the quality of evidence we need to announce something so monumental as the discovery of alien life.

You'd think, given what's at stake, that NASA would have repeated the LR experiment on subsequent missions, right? Wrong. Since 1976 *none* of the orbiters, landers, and rovers sent to Mars have been fitted with life-detection experiments. It may be that NASA is too risk averse, or doesn't want to be associated with what would be a perceived failure if it did not detect microbial life. Perhaps. More likely, as we'll see, it's because the detection of life is something that seems like it ought to be simple, but turns out to be immensely difficult.

Such are the considerations and compromises that must be made when asking Congress for funding. This is where we come in. The cost of the Viking mission was about $1 billion in 1970s dollars,[13] or $5 billion today. It would cost at least that to send a rover, which we could do, although as it happens, ESA has one on the way.

ESA's mission, in collaboration with the Russian space agency Roscosmos, is part of its ExoMars program—its title standing for "exobiology on Mars." The first part of the program, a planetary orbiter, has already launched and is making measurements of the atmosphere. The second part contains a rover (named after Rosalind Franklin), and delays mean it's not due to launch until 2022. It is designed to look for signs of life on the planet, with cool bits of equipment on board such as the MicrOmega,[14] an infrared "hyperspectral microscope" that will be able to closely analyze the composition of soil samples, and a gas chromatograph to analyze soil chemistry.

The most important instrument is the Mars Organics Molecule Analyzer (MOMA[15]), which, crucially, can determine whether molecules are left-handed or right-handed. Many organic molecules can exist in one of two chiral forms—essentially they are left- or right-handed in terms of their molecular structure. Sometimes, organic compounds such as methane are made by nonbiological processes such as geological activity or erosion, and when this happens and we measure the compounds we find an equal number of left- and right-hand forms. But biology prefers its molecules a particular way. The sugars used by life-forms are right-handed, and the amino acids are left-handed—no one really knows why. So a chirality analysis is an essential test to perform on organic compounds. If *Rosalind Franklin* finds a bias toward one kind of chirality, it is powerful evidence that there are microbes living on Mars.

If that's the case, we may also be able to tell if the life-forms are similar to ones on Earth, or if they evolved after a separate origin-of-life event.

In the distant past, Earth and Mars were exchanging rocks all the time in the form of meteorites dislodged by asteroid impacts—an estimated one billion tons going each way—and it's quite possible that some bacteria hitched a ride one way or another. Back in the day, Mars had a magnetic field so was shielded from radiation, and the planet seems to have been warmer, with liquid water, even if there were periods of glaciation billions of years ago.[16] The NASA rover, *Perseverance*, which arrived on Mars in 2021, will investigate. Even now, the subsurface rocks of Mars are porous, which make them more suitable than Terran rocks for microbial life, because there are more opportunities for life to squirrel itself away.

It could turn out that effectively we're Martians. Or it might be that life arose independently, which would perhaps be even more thrilling than finding out that our ancestral home is Mars. Both the orbiter and the rover cost ESA €1.3 billion; NASA's *Perseverance* rover is budgeted at $2.7 billion. We need to get another couple of rovers up there. Craig Venter, one of the driving forces behind the sequencing of the human genome, and whose research institute leads the way in building synthetic life-forms, once said he wanted to send a DNA sequencer to Mars. We should certainly make this happen. We might have to do a sample return mission: Have a rover dig up samples, then have another spacecraft collect and return them to Earth for sophisticated analysis. This is part of the *Perseverance* plan.

My primary interest is intrinsic. It's about the thrill of finding an actual alien life-form. No matter that it is most likely bacterial. But we also need to survey the planet properly before billionaire entrepreneurs start crashing around up there. The contamination of another planet that may harbor living things, or even fossils of past life, is something we have to be very careful about.

～

JUPITER HAS SEVENTY-NINE MOONS, and the four biggest were discovered by Galileo in 1610. Galileo solicitously named them after the four Medici children in the hope of getting funding—for scientists, it's the same old story—but his names didn't take, and the moons ended up being known by the names of some of the lovers of the god Jupiter: Io, Europa, Ganymede, and Callisto.

The one we're most interested in is the second closest to the giant planet, Europa, which I see regularly with my telescope from my bedroom window. It is only about the size of our Moon, but is shiny with ice. This is why we like it: Deep below the ice is an ocean of liquid water, and water is key to life. The ice is miles thick, which means it will be tricky to drill through, but fortunately it looks like geology helps us out. The Hubble Space Telescope spotted great gouts of water jetting up from the surface of Europa, reaching up to 125 miles above the surface. We could sample the plumes directly from orbit, or we can land on the Moon's surface and dig around in the ice. Recent work shows that even a thick layer of ice is enough to protect any microbes from the intense radiation at the surface.[17]

NASA and ESA have Europa missions in the works. ESA has the Jupiter icy moons explorer (Juice) mission, and NASA has the *Europa Clipper*. Both are flyby missions and plume chasers. The idea is that the spacecraft can locate a plume and adjust course so as to fly through it, catching samples of ice for analysis. But, even if this is successful, the results are not going to tell us if the subsurface ocean contains microbes, as the instruments on the probes are designed to check for characteristics *associated* with life, not life itself. The problem of designing a life detector is one that plagues space agencies, and with good reason. It's mighty difficult to build an experiment or set of experiments that can proceed entirely autonomously in the harsh conditions of space, and that can reliably and unambiguously demonstrate the existence of life.

One of the issues scientists have is that there is only a working definition of what life is. NASA's is that it is a "self-sustaining chemical system capable of Darwinian evolution." But finding ways to test for that is the problem. After the Viking mission, as we've seen, NASA has not even attempted it. The agency is, at least, trying to find ways to do it, and has created what it calls a Ladder of Life Detection,[18] a summary of the different features of life that *could* be detected, along with assessments of the likelihood of being able to disprove the hypothesis that the sample is in fact nonbiological.[19] The bottom rung of the ladder is "habitability," which is essentially the mere presence of liquid water, moving up through the detection of various biomolecules, including DNA and RNA, metabolism, reproduction and, finally, Darwinian evolution. Tests of this last rung are unlikely to be possible by a robot mission any time soon, and even the other rungs have problems— there is no simple and unambiguous test for life.

Still, happily we don't have to make a case to Congress. We can just sign off a small amount of our budget, say 1 percent, to go toward Europa missions. Don't hold your breath: Developing the spacecraft will take a few years even at top speed, and getting to the Jupiter system would take another three years.

⌐⌐

DESPITE LONDON'S LIGHT POLLUTION, I can sometimes also see Titan from my bedroom window—though it isn't much to look at. Probably just about the same as when Christiaan Huygens discovered it in 1655: a pinprick of light next to Saturn, which itself, with my telescope, looks like a tiny dot with ears. Get a bit closer, however, and you'll see why Titan has long been one of the most exciting locations in the solar system.

First, it has a dense atmosphere, the only moon in the solar system to have one. The air is around 97 percent nitrogen and

3 percent methane, with a bit of hydrogen and a few other trace gases. It also has weather systems: Clouds of methane in the atmosphere periodically rain, and lakes and rivers of liquid methane have been seen on the surface. Other than Earth, it's the only place we know of in the solar system with this kind of surface feature and this kind of cycling of compounds.

Rather than a water cycle like we have, Titan has a hydrocarbon cycle where methane and ethane are rained down and eventually cycled back into the atmosphere.[20] But Titan is very cold, with a surface temperature of −290°F, making surface living difficult, to say the least. Imaginative scientists have come up with plausible kinds of life-forms that could breathe hydrogen instead of oxygen, but a membrane-bound life-form—basically, life as we know it—doesn't look possible. Under the surface it's warmer, and there is a vast ocean of liquid water. It is here where scientists are more hopeful that life could thrive. NASA has a mission to Titan in the works—that was supposed to launch in 2026 and to reach Titan in 2034—but it has been delayed because coronavirus has eaten away at sources of funding. The probe is a plutonium-powered drone quadcopter called Dragonfly, and is budgeted at $850 million.

Several other missions have been proposed, but none yet have a green light. They include a joint NASA/ESA exploration of Titan and other moons of Saturn, a reconnaissance drone designed to make a high-resolution map of the surface, and an intriguing plan to send a submarine into Titan's freezing hydrocarbon oceans.[21]

The *Titan Turtle* is a nearly seven-foot-long, 1,100-pound sub that would be delivered by a spacecraft and dropped by parachute into the sea, where it would keep warm and explore with its plutonium-powered electric engine. There's a lot that needs working out, but Steve Oleson of NASA's Glenn Research Center in Ohio has put together a plan. Internally the craft will be warmed by the plutonium generator, but the external surfaces of the sub will have to be able to work in liquid methane. The seas of Titan are about 650 feet deep, which in a medium of liquid hydrocarbons is only about a pressure of two Earth atmospheres, so the sub would be able to take sediment samples from the bottom. This is such a (literally) cool idea, we should make it happen.

~

ENCELADUS, NAMED AFTER ONE OF THE GREEK GIANTS (the son of Gaia, the god of Earth, as it happens), is only one tenth the size of its neighboring moon Titan but, being covered in fresh ice, is far brighter. It is this ice, and specifically the brightness of it, that makes it interesting. The reflectivity of the ice makes the Moon colder than Titan, and the surface temperature is –324°F. Enceladus has a subsurface ocean of liquid water, but how was the surface staying so bright and not getting scuffed and dulled by erosion?

When the Cassini–Huygens mission to Saturn started surveying the system in 2004, the spacecraft detected something odd in the atmosphere of Enceladus. Michele Dougherty from Imperial College London was the scientist in charge of the magnetometer on Cassini and, convinced by the data she was seeing, lobbied NASA to divert the spacecraft to make a low flyby over Enceladus. Making an adjustment to a deep-space flight plan is a big deal, as every drop of fuel is accounted for, and any change of course means spending less time on another part of the mission. But Dougherty successfully made her case as to why it was worth taking a closer look at Enceladus, NASA made the course adjustment and the risk paid off.

Cassini ended up passing just 107 miles over the surface of the moon, and made a spectacular discovery: Enceladus was emitting plumes of water into space. This was *very* exciting. This explained why the surface was so shiny and new, because it was being regularly covered in fresh ice. And astrobiologists immediately understood that the conditions were compatible with life. There are microbes on Earth that live by metabolizing hydrogen and carbon dioxide to produce methane, and these have been subjected to conditions similar to that on the subsurface of Enceladus, and survived.[22] Tantalizingly, organic compounds have been detected in the plumes.[23] The oceans of Enceladus could be teeming with alien life.

Enceladus, in short, is very interesting and worth a trip. A team from NASA's legendary Jet Propulsion Laboratory at Caltech in Pasadena does have a proposal, the Enceladus Life Finder mission, which they say is the natural follow-up mission to Cassini, but the agency declined to fund it in the last round of applications. There has been some talk of a mission that would be privately funded by Yuri Milner, the billionaire entrepreneur behind the Breakthrough Starshot project, which aims to address deep scientific and technological questions (a bit like some of the questions in this book, but with less cash). Milner started his career as a physicist before becoming an investment banker, and was an early investor in Facebook. It's good to have the odd billionaire interested in science rather than yachts and golf courses. NASA has agreed to support Milner's proposal, to the initial tune of $70,000, which is hardly going to get us out of California, but it's a start. We will immediately fund both the Milner and the Enceladus Life Finder enterprises.

∾

AS WELL AS THESE ROBOTIC SEARCHES for life based in our own solar system, we can also search on exoplanets: worlds in other systems. There is an ascending scale of reasonableness to the ways we can do this, starting with land-based telescopes, then moving into space and on the Moon, then to the more ambitious—some might say fantastical—plans to send robot spacecraft across the vast distances of interstellar space. Finally, we'll look at NASA's long-term plans for human missions to other star systems.

In Hawaiian mythology, the island of Hawaii is the child of the father of the sky, Wakea, and the mother of the Earth, Papahanaumoku. The mountain of Mauna Kea, rising more than thirteen thousand feet above sea level, is the conduit that connects the island and its people to the heavens, and as such has a sacred place in Hawaiian culture. Its altitude and remote position in the Pacific also make it ideal for astronomy, and there are thirteen telescopes built on the summit. A fourteenth, and the biggest so far, is currently being constructed—or it was until protests stalled the project. The Thirty Meter Telescope (TMT) will cost around $1.4 billion, and when completed will be by far the largest visible-light telescope in the world. It will open new pathways in the investigations of black holes and the structure and evolution of the universe, and the ability to analyze the characteristics of exoplanets.

There has been significant controversy over the project, with some seeing it as further desecration of the sacred site. Critics say construction projects must incorporate *aloha 'Aina*, the Hawaiian love of the land, and acknowledge the cultural value of the location. If the impasse is not resolved, the project may have to be completely relocated, perhaps to a site in the Canary Islands.

We will need to be aware of this for our project: building a much larger array of optical telescopes on the far side of the Moon. The Moon has huge cultural significance, and people are going to be upset at the thought of one of the biggest construction projects in human history going on up there. Hopefully the fact that our site is on the far side will help. A telescope there, while much more expensive to construct than on Mauna Kea, is best placed for observations, because the Moon itself shields the telescope from Earthly radio interference. NASA has no current plans signed off for a lunar telescope. We have the opportunity to make this happen.

If we *did* build lots of similar telescopes on the Moon, the absence of light and radio pollution would give extraordinary resolution. "Spread out over a few thousand miles of lunar landscape, such an instrument would be able to see details as small as the city of Los Angeles from one hundred light-years away," says Seth Shostak, the senior astronomer at SETI in California. "If aliens have built large constructions in the vicinity of a nearby star, this instrument would have a good shot at seeing them."

Building a single thirty-meter telescope in Hawaii is a big job; doing it multiple times on the Moon is considerably more demanding, requiring a permanent Moon base, the development of human-capable rocket transport and multiple trips up there during construction. But, as we saw in the previous chapter, it can be done within our budget.

Nor should we content ourselves only with optical telescopes. The Square Kilometre Array (SKA) is a radio telescope planned for remote regions of South Africa and Australia. The facility will have radio antennas spaced over thousands of miles and will combine their signals to simulate one gigantic radio telescope. Construction was supposed to start in 2019, but inevitably delays and cost-hikes have stalled the project. It is currently projected to be complete by 2028, at a cost of around €2 billion.

The Array will enable tests of general relativity, an investigation into the effects of dark energy, and will allow astronomers to observe "first light." This is the time, starting 380,000 years after the big bang, when stars first formed from the hot gases of the early universe and the first light shone in the cosmos. The SKA will be able to observe the period when the first stars and galaxies formed.

It will also allow us to look in detail at the planets around nearby stars in habitable zones, the region where temperatures are just right for water to exist in liquid form, and so be most conducive to life as we know it. The atmosphere of these exoplanets can be examined for chemical signs of life. And, if there are alien civilizations watching TV on exoplanets near us, the SKA will be sensitive enough to pick up the transmissions.

As we've seen, telescopes off-planet are better, because they are free of interference from Earth's atmosphere. There is the atmosphere itself, then there are the radio waves that fill it from all our transmissions and TV shows and mobile phones, but then there are the soon-to-be-thousands of new satellites going into orbit. SpaceX is building the Starlink satellite constellation, a network of tens of thousands of small satellites launched into low orbit to provide internet access from space. (Some scientists are worried that they could ruin astronomy from Earth.)

For all these reasons, we need to get into space, and we'll sign off on optical and radio telescopes for the Moon.

There are great examples of what we can get when we leave Earth. The Kepler space telescope, which NASA retired when it ran out of fuel in 2018, discovered around 2,600 alien worlds, with as many as 3,000 more "candidate" planets awaiting confirmation of their existence. If you're keeping count, Kepler cost $700 million.

One problem with Kepler was that it stared at the same patch of the universe, and many of the planets it found are very far away, even by galactic standards. Its replacement, the TESS satellite, is tasked with looking for exoplanets closer to Earth. Once TESS (the acronym stands for "Transiting Exoplanet Survey Satellite") has found these planets, we can train different instruments on them to measure their mass and atmosphere more closely.

One of these instruments will be the James Webb Space Telescope, which has also been massively delayed. The problem is not so much the technical difficulty of making the thing as ballooning costs. It is now projected to cost $9.66 billion. Had we been responsible, obviously, these problems wouldn't have affected us. The James Webb, a collaboration between NASA and the Canadian and European space agencies, is the successor to the Hubble Space Telescope, and will have a mirror five times larger than Hubble's. It will eventually sit at a stable Lagrange point, a point between Earth and the Sun where the gravity from each cancels out, and will enable us to image the atmosphere of exoplanets in better detail than ever before. We may even be able to determine if there are gases of biological origin in exo-atmospheres.

Our job is to accelerate current projects, such as the TMT, the SKA, and the James Webb, and to initiate a more ambitious Moon-based array.

~

Moving from the search for microbial aliens and habitable exoplanets, we turn now to the holy grail and, in the public mind at least, the ultimate glory: the hunt for alien civilizations. This is the mission of the SETI Institute—the search for extraterrestrial intelligence, started by astronomer Frank Drake in the 1960s. Drake is famous for his attempt to estimate the number of alien civilizations in our galaxy that might be broadcasting radio signals at any given time. His lower estimate is twenty, the upper estimate is fifty million, and, despite criticism over the years, just the merest chance that we might hear a signal from an alien civilization has captured the imagination of millions. SETI has spent six decades listening for signals in various ways and using various radio telescopes.

Only once has it heard something significant. On August 15, 1977, Jerry Ehman, an astronomer working at Big Ear, Ohio State University's radio telescope in Delaware, picked up an incredible signal from 220 light-years away, in the direction of Sagittarius. He wrote "Wow!" on the printout from the telescope, and the burst of radio transmission has since been known among aficionados as the Wow! signal.

There are two ideas for what caused the Wow! signal. The telescope was tuned to a radio frequency of 1,420 MHz, which is the wavelength of hydrogen, the dominant component of the universe. Astronomers spend a lot of time looking at this wavelength, because it reveals the most information about the structure of the galaxy. So one idea is that the blip was registering a surge in hydrogen, maybe from a comet. The other is that it was an interstellar beacon, a signal sent by an alien civilization: 1,420 MHz is also the frequency that alien astronomers will most likely be using, if they're out there. But the signal, or the blip, has never been seen again and we're unlikely to ever know what it really was.

The problem for SETI is getting access to valuable time on radio telescopes, and we can at least help with that.

There is a faction of alien hunters who try to make the hunt more active than passive. Doug Vakoch is director of Interstellar Message Composition at METI in San Francisco. The acronym this time stands for "messaging extraterrestrial intelligence." Instead of sitting and listening and searching for signals *from* aliens, he says, we should actively direct signals *to* them, or to promising exoplanet locations where aliens might live.

The idea is to use a large radio telescope to send messages to stars within eighty-two light-years of Earth. This has led to much debate among scientists. What we might call the Borg argument, made against Vakoch's plan, goes that if we alert a super-intelligent species of alien to our presence, they'll come here and harvest us, like the Borg in *Star Trek*. This fear motivated several high-profile scientists, including Stephen Hawking, to warn against shouting about our location to aliens. In a call for a moratorium on METI, the worried faction said: "Intentionally signaling other civilizations in the Milky Way Galaxy raises concerns from all the people of Earth, about both the message and the consequences of contact. A worldwide scientific, political and humanitarian discussion must occur before any message is sent."

Or, as Hawking once put it, alien contact might put us in a similar position to the Native Americans at the point when Europeans first arrived, "which didn't turn out very well for the Native Americans." Only Hawking could get away with that sort of remark, but it was a daft point to make, really, since aliens are hardly going to make journeys of multiple light-years just to collect something from Earth—iron or carbon or whatever. Any advanced alien civilization capable of interstellar travel will be able to make what it wants from the natural resources of space. Unless they really are like the Borg and are driven for some reason to assimilate all galactic cultures.

It's conceivable that malicious aliens could harm us from afar, by sending messages at light speed. This is the basis of *A for Andromeda*, by astrophysicist Fred Hoyle (he coined the term "big bang" in derision about the concept) in collaboration with the writer John Elliot. Similarly, Cixin Liu's *Three-Body Problem* epic starts when aliens from the Alpha Centauri system make remote contact with humans on Earth. When I spoke with Liu about searching for extraterrestrials, he did warn of the danger of initiating a crisis. "We are not an advanced civilization," he said. "We are like toddlers, and even toddlers can cause big problems."

The hunt for extraterrestrial intelligence is further advanced in China, by the way, than in the US. Arecibo used to be the world's largest radio telescope, but that title now belongs to the Five-hundred-meter Aperture Spherical Radio Telescope, abbreviated as FAST and nicknamed Tianyan ("eye of heaven"), in the mountains of southwest China. FAST cost $180 million to construct,

but $270 million in relocation costs and payouts to the 9,000 peo-
ple relocated when their village was destroyed to make way for it.

⌒

GIVEN THAT—according to Drake's back-of-the-envelope
calculations—there are at least twenty advanced alien civilizations out there, and maybe millions, but that we haven't heard
from any of them, the obvious question, asked first by physicist
Enrico Fermi, is "where is everybody?"

One answer may be the Great Filter. This is the idea that a gigantic challenge faces every life-form during its evolution that
filters out—kills off, in other words—its chances of reaching a
galaxy-spanning level of development. It might be that it is incredibly unlikely for microbial life to make it to even a multicellular level. Or it might be that nuclear war—or climate change—
wipes out intelligent civilizations before they have time to reach
out to the stars. It's a fun thing to consider but hard to test.

More likely than malicious aliens, to my mind, is that there
are aliens out there who are *not* sending radio signals.* Sara Seager, the MIT professor who is part of the Venus team, revised the
Drake equation to take this into account, providing an estimate
of the number of planets with detectable signs of life that it might
be possible to discover in the near future. Her equation predicts
that an atmosphere on an exoplanet showing signs of life could be
discovered twice in the next decade.[24]

A promising place to look is the planetary system around a
star called GJ 887, 10.7 light-years away, which has at least two

---

* By "aliens," here, I mean microbial life, not necessarily intelligent life.
  Although Cixin Liu thinks that if there is intelligent alien life out there, it will
  be keeping silent. His argument, which he calls the Dark-Forest theory, is that
  advanced civilizations remain silent, like hunters in a forest, so as not to give
  away their position to other hunters who will seek to destroy them.

(and possibly three) rocky, Earth-like planets. Encouragingly, GJ 887 seems less prone than most stars in our galaxy to flaring up in atmosphere-destroying tantrums, meaning its planets might have had enough time to evolve life. Another update to Drake's equation adjusts it to account for the uncertainties around the different parameters, such as the likelihood of life evolving. The course-corrected Drake equation now finds that the chance we are alone in the universe is high: 30 percent. Fermi's question—where are all the aliens?—is sometimes called a paradox, but the paradox dissolves substantially with this more rigorous statistical approach. Intelligent aliens may well not exist.[25]

We'll bear that in mind as we turn to more ambitious plans to search for alien life.

Cixin Liu had his aliens originate from the nearest star system to us, Alpha Centauri, just over four light-years away. That's 25 trillion miles. It would take tens of thousands of years for us to reach it with our best rockets. Which is where Yuri Milner, the venture capitalist we met on the way to Enceladus, has other ideas. Milner is serious about space exploration, and in 2016 announced Starshot, a project to research the feasibility of sending a laser-powered spacecraft to Alpha Centauri. Yes, we just said it's more than four light-years away, but the idea is that we accelerate spacecraft to 20 percent the speed of light, which means they would get there in twenty years.

The proposal, backed by $100 million of Milner's money, is to construct a light sail that is pushed by an ultra-powerful laser generated from a device based on Earth, or perhaps in space. The sail, a few meters wide, is attached to a tiny spacecraft that weighs only a few grams, and the thrust delivered by the laser would be enough to lift the spacecraft to 0.2 light speed (it would flash past Mars in twenty minutes, Pluto in seven hours). None of the technology has yet been demonstrated. Milner has some tentative support from scientists, although it's hard to say whether they are genuinely convinced the project has merit, or if it's more that they are happy to use the money for interesting research.

There are multitudes of technical and political hurdles to overcome. The light sail has to be strong enough to withstand the blast of a laser sending it to immense speed, and it has to reflect almost all of the laser, so as not to overheat. Materials scientists must make something out of silica or diamond that is thinner than spider silk but more reflective than anything we know, and create a sail out of it that weighs a gram or less. Unless we can utilize the laser heat somehow, we also need a power source on the spacecraft; plus thrusters for course correction, a transmitter and receiver that will work across four light-years, and a camera and perhaps some other scientific instruments. The lasers themselves will require sixty gigawatts, the power of twenty standard nuclear power stations, and will effectively be weapons of mass destruction. It will be a political nightmare trying to get governments to agree to its construction.

The Starshot team think the total cost will come in at around the price of the Large Hadron Collider: $14 billion. That seems wildly optimistic. I think we should provide some funding, however. The research alone will generate a lot of useful spin-off information.

Finally, if a miniature spacecraft traveling at relativistic speed is ambitious, far more so is the 100 Year Starship. This is a project initiated by NASA and led by Mae Jemison, a chemical engineer and former astronaut (she was the first black woman to travel in space). The project is designed to ensure that in the next 100 years humans will have the capability to travel to other solar systems. Jemison believes that we would be on Mars by now if the political desire for space travel hadn't dried up. Of the many problems we need to solve, however, a fundamental one is how you power a spaceship across interstellar distances.

Antimatter propulsion is one of the candidate power sources that may eventually be used in future space travel, but we are very far away from being able to get or make enough antimatter. Antimatter produces vast amounts of energy when it annihilates in contact with regular matter, but we can only make minute amounts of it. It's probably worth sending a couple of dedicated antimatter-hunting spacecraft to look for the stuff. We could search in the Van Allen radiation belt around Earth, which has small amounts of antimatter, and a similar one around Jupiter, which may have more.

Another type of engine to develop is the fusion rocket. This first requires us to figure out how to recreate nuclear fusion processes safely on Earth (which we looked at in chapter 3). Once we've done that, which will cost tens of billions of dollars, we need to miniaturize the technology and incorporate it into a spacecraft. The big advantage of a fusion rocket is that we wouldn't need to carry the immense amounts of fuel we rely on for conventional rockets, and it allows us to accelerate over long distances so we can build up great speed.

More primitively, we could explode nuclear bombs behind a spacecraft and ride the propulsive wave. This was the basis of Project Orion in the 1950s, which the physicist Freeman Dyson worked on for a year until everyone realized it wasn't going to be diplomatically feasible.

IF WE ARE TO FIND ALIEN LIFE of any form, and if we are to explore our cosmic neighborhood properly, we need to work on a range of these projects. The desire to look outward, to explore, to understand the environment, is fundamental to human nature. Our environment now includes the planets and moons in our solar system, and the planets and stars within ten light-years of us.

Let's get out there. Let's bring an end to cosmic loneliness. I was startled to find that it is so cheap, certainly within the scope of this book, to send robot craft on detailed exploratory missions all across the solar system, and to probe the deeper galaxy with telescopes. In terms of public outreach value, it may be the mega project that has the best cost-to-inspiration ratio.

Achieved

We don't know what we haven't yet discovered, but with an extensive enough search I'm 70 percent convinced that we'd find simple life-forms in our solar system. Even without this, our missions will generate massive amounts of information about our immediate and wider galactic neighborhood, and deliver huge spin-off benefits in the form of innovation and scientific discovery.

Money spent

In-depth robot missions to Venus, Mars, Europa, Titan, Enceladus:	$150 billion
Construction of Earth, space, and Moon-based telescopes:	$30 billion
Development of interstellar spacecraft:	$50 billion
Total:	**$230 billion**

Edvard Munch's *Scream of Nature*, painted in the wake of the Krakatoa explosion, which left the skies red across Europe

Redesign Our Planet

AIM: To remove carbon dioxide from the atmosphere such that we eventually return to a safer concentration of 350 parts per million. To buy us time to decarbonize the global economy and keep global heating to less-than-catastrophic levels.

I USED TO THINK THAT *The Scream* by Edvard Munch represented a person overwhelmed by some internal mental horror. Probably it does. But now it makes me think of climate change. What I didn't know was that Munch titled the piece *Der Schrei der Natur* (*The Scream of Nature*), and the explicit link to the planet in the German title is not accidental. The unforgettable blood-red sky that Munch painted in the background was his memory, perhaps, of the atmospheric aftereffects of the eruption of Krakatoa, in 1883, which were visible even in his native Norway.[1] When the volcano exploded, vast amounts of sulfur dioxide were thrown into the atmosphere. The gas spread around the world, changing the reflectivity of the clouds and diverting sunlight from the planet. Spectacular and eerie sunsets occurred around the world, and the following summer the temperature in the northern hemisphere fell by around 1.2°C (roughly 2.2°F). That, coincidentally, is almost the amount we've now increased the average global temperature by since preindustrial times. What we would give now to have a lasting decrease in average temperatures of 1.2°C (2.2°F)! Well, let's find out. What *would* we give?

Our question, and it's a massive one, is this: Could we mimic the effect of Krakatoa? All we'd need to do is heft a lot of sulfate particles into the sky, by balloon or plane. It might not even cost much of our money. We could fix global warming. We could cool the planet.

That's the idea, anyway, and it's a seductive one for people looking for a technological fix for climate change. But geoengineering is an idea fraught with danger and uncertainty. When Krakatoa cooled the planet, it also skewed weather systems around the world. Southern California had a year of record rainfall; other regions had drought, and crops were lost. When Mount Pinatubo erupted in 1991 in the Philippines, we were able to measure the impact on the atmosphere more accurately. The explosion injected around 19 million tons of sulfur dioxide into the sky, which, as with Krakatoa, increased the reflectivity of the clouds, this time by about 10 percent. Temperatures in the northern hemisphere fell as result by 0.5° to 0.6°C (0.9° to 1.1°F); globally the fall was 0.4°C (0.7°F). But ozone depletion spiked as a result, and the eruption may have contributed to America's "storm of the century" in 1993.

If we want to interfere with the Earth system on this scale, it's not enough to ensure that any attempt does more good than harm—we have to compensate those people adversely affected. Any budget will factor in these kinds of payments to the total cost. But we may need to try some kind of geoengineering scheme—or at least to prepare the way, to have the possibility of setting one in motion, if (or when) it is essential. In 2019, Johan

Rockström of the Potsdam Institute for Climate Impact Research in Germany said that the climate emergency was so severe that we should consider geoengineering.[2]

It is in some sense already too late. A rise in sea level is already locked in, while the decline of many mountain glaciers and two huge glaciers in western Antarctica has passed the point of no return—they will melt, even if we stopped carbon emissions tomorrow.[3] The same may be true for the massive Greenland ice sheet, the second largest ice sheet in the world, after the one that covers Antarctica.[4] And there is so much ice on Greenland that if it all melted, it would raise sea levels around the world by twenty feet.

In 2020, heat records were set across the planet, from Japan to California. The relentless heat in the southwest United States threatens a long-lasting megadrought, while heat-induced humidity in East Asia is starting to approach the limits of human survivability. There was intense flooding in India, and in China, threatening the massive Three Gorges Dam, the world's largest hydroelectric facility; plus out-of-control wildfires in the Arctic and in California, unusual hurricane activity—we're barely coping with a 1°C (1.8°F) temperature increase. But we are starting to get an idea of what we might experience with 1.5°C (2.7°F) of warming. At 2°C (3.6°F) or more, much of the planet would become intolerable for human life.[5] And remember—we're currently heading for a world that is 3°C (5.4°F) warmer by the end of the century.

The clamor for some sort of technological solution may eventually become irresistible to politicians. Perhaps it will be the fear of sea level rise, as hundreds of millions of people in coastal megacities seek help. Not to mention the trillions of dollars tied up in those places. The Institute for Economics & Peace produced an Ecological Threat Register in 2020, and found that 1.2 billion people in thirty-one countries could be displaced due to climate-change-related problems such as drought or food short-ages by 2050.[6] As these sorts of problems and impacts mount, we have to be prepared for a pivot from inaction on emissions to action on an engineering solution.

Even now, we're teetering on the edge, so in this chapter we're going to see if we can buy time while we move to zero carbon. We'll look at the planetary prognosis, at what's at stake if we don't limit global warming. Then we'll look at what we could do with our money to physically cool the planet.

⌒

THE IDEA TO MANIPULATE THE EARTH SYSTEM has been around for about as long as the knowledge that the planet is warming. In 1965, when US president Lyndon Johnson's Science Advisory Committee flagged the dangers of greenhouse gas emissions, they didn't recommend reducing fossil fuel use. They instead suggested increasing the reflectivity of the Earth to offset the heating. In other words, geoengineering. The Convention on Biological Diversity defines this as "a deliberate intervention in the planetary environment of a nature and scale intended to counteract anthropogenic climate change and/or its impacts." Now, we're going to look at three different kinds of engineering solutions.

First, ideas inspired by Krakatoa. "Solar geoengineering" is the term given to proposals to screen out some of the sunlight reaching the surface of the planet. Some people call this solar radiation management (SRM), which may give the impression that manipulating the atmosphere is simply a case of adjusting the dimmer switch on the Sun. In fact, and we need to be clear about this from the start, we don't even know if it will work yet, and we don't know what the side effects might be. It's promising, for sure, but we're a long way from being confident in it. One major problem with solar geoengineering is that it will cool the tropics more than the poles, so it will be of limited help with curbing sea level rise.[7]

Second, we'll look at ideas to suck out carbon dioxide from the atmosphere, reducing the concentration of greenhouse gases and thereby reducing the greenhouse effect. This itself comes in three flavors: a biological or natural kind, which usually means planting more trees, or growing more plants in the ocean; a technological kind, which means extracting carbon dioxide from the air and burying it underground forever; and a hybrid techno-natural way, which means growing plants for fuel and then catching and burying the carbon dioxide produced when the fuel is burned.

As with solar geoengineering, we don't yet know if this "negative emissions" technology of carbon dioxide removal will work well enough at scale, or be cost-effective. However, just the *idea* that negative emissions might one day work on a significant scale justifies the Intergovernmental Panel on Climate Change (IPCC) in its claim that we may be able to limit warming to 2° or even 1.5°C (about 3.5° or 2.7°F). As well as the massive emissions cuts that the IPCC models say we need to make, we also need to start capturing and storing carbon dioxide in vast amounts.

Finally, we'll look briefly at how we could engineer the economy in order to cool the planet. We should acknowledge the extreme sensitivity around this topic, and particularly around solar radiation management. Most climate scientists shudder at the thought. One leading *advocate* of geoengineering angrily told me it was "illegitimate" to spend my money on solar radiation management and that it would be the precise opposite of good governance. Even many of those who work on it say that the best use of geoengineering would be to find a way to avoid doing it and, failing that, we should use it only when it is tied to reductions in carbon emissions (and limit it to removing what we've already put into the atmosphere).

As we'll see, there is a lot of interest among billionaires in various geoengineering schemes, but something that affects the entire planet must be something regulated and governed democratically, and we can tie that sort of oversight into our spending. Critics say that proponents of geoengineering are enablers; that it will empower us to carry on burning fossil fuels and that it won't force us to make the emissions cuts that are required. We mustn't fall into that trap. It's the same with proponents of carbon capture and storage (CCS): People fear that the lure of this unproven technology will prevent us from making emissions reductions.

Not that we're making them, anyway. And geoengineering might instead encourage emissions cuts. "Maybe intervention would be positive," said Cecilia Bitz, a sea-ice physicist at the University of Washington, Seattle, "showing that we have the capacity to improve the environment." A study in 2015[8] found that just learning about geoengineering increased public concern around climate change.

⌒

BEFORE WE LAUNCH INTO THE TECH, let's be clear why it's still essential to make emissions cuts. Imagine we did put up a sunshield to cool the planet. We do not want to commit to that being permanent, we just want one up for one hundred years or so, to give us time to cut and then halt emissions and get rid of a lot of the carbon dioxide already out there. If our sunshield comes down before we've removed the carbon dioxide and other greenhouse gases, the temperature would leap back up, so we will need to phase out our shield only when we've removed sufficient gas from the atmosphere for the greenhouse to cool down.

A key point about climate change is that it isn't just about the carbon we put into the atmosphere, it's also about the amount that is constantly being drawn down into the oceans, soils, and biomass. About a quarter of the carbon dioxide that goes into the atmosphere from our activities gets absorbed into the ocean and becomes carbonic acid. The ocean is vast, but so is the amount of carbon dioxide we've been putting out, and the rate of acidification is far too rapid for sea life to adjust. Screening out sunlight alone won't tackle that problem.

There are many nightmare scenarios, so let's just pick one. About twelve thousand years ago, in what is now the Barents Sea between Norway and Russia, there was a series of more than one hundred large explosions on the ocean floor. Pockets of methane that had been trapped for thousands of years under the seabed bubbled up into the atmosphere. We know this, because a study in 2017 revealed craters on the seabed left behind after the explosive release of gas. There are still huge amounts of methane, a greenhouse gas twenty-five times more potent than carbon dioxide, trapped under the seabed, especially in the Arctic. It is held in a slushy frozen form known as clathrate. But, as the ocean warms, the gas is starting to bubble out. The Arctic is heating up faster than anywhere else on the planet, and between 2007 and 2016 the Siberian permafrost warmed by a full degree Celsius (nearly two degrees Fahrenheit). It's one reason the wildfires in the Arctic in 2019 and 2020 were so bad.

Methane levels have been rising gradually over the last decade or so, but we don't yet know if this is from clathrates becoming unstable as the world gets warmer or from the general background increase in methane as a result of human activity. It may be the latter. No one knows how likely it is that we will get an explosive release of methane, as we did twelve thousand years ago, but if it comes out into the atmosphere in a burst rather than a trickle, it doesn't really matter even if we fully implement our pledges on emission reductions: The effect of the methane will wipe out all our gains. We could face a sudden, dizzying, and nightmarish jump in warming.

If that happened, the pressure to attempt a geoengineering fix may be irresistible. It is essential we make plans so we can deal with any emergency with some degree of preparedness. But even without the methane explosion scenario, we desperately need to do something to slow the rate of warming.

The amount of carbon dioxide in the atmosphere is measured in parts per million (ppm). If we carry on burning fossil fuels at the current rate, we would go from around 415 ppm, where we are now, to around 1,200 ppm at the end of the century. That will cause 4°C (7.2°F) of warming in the IPCC's conservative assessment. However, a review of thirty-nine different models of climate change found we could get to 4°C (7.2°F) as early as 2065.[9] One computer model suggests that, if we get to 1,200 ppm, cloud formation will be disrupted over large parts of the subtropics, potentially adding another 8°C (14.4°F) of warming on top of what we've already got.

The question, and it's a critically important one, is: Can we engineer the climate responsibly? The US National Academy of Sciences doesn't like the term because it points out (stuffily, but correctly) that engineering only works on systems that are understood, and we don't properly understand the climate. So we will commit to invest in research to properly understand it.

The Academy also doesn't talk about geoengineering as one subject, splitting it into two in the same way we have and talking about reflecting sunlight, and removing carbon dioxide.

∽

The most dramatic forms of geoengineering are methods that attempt to change the reflectivity, or albedo, of the planet. The National Academy of Sciences prefers to talk about albedo modification rather than solar radiation management, again because "management" implies it is something that can be done with some degree of certainty. The ideas for this range from putting a giant mirror in space to reflect sunlight (which is beyond even our money), to blocking some of the sunlight getting through to the earth's atmosphere, through some kind of sunshade, mimicking volcanic eruptions that darken the skies with ash.

The sunshade is perhaps the most promising. As currently envisioned, it is very far off from doing anything on the scale of Krakatoa. In most scenarios, aircraft fly high in the atmosphere, releasing a steady stream of sulfate particles that (while they remain in the air) block sunlight from reaching the ground.

The most advanced plans to date are those devised by David Keith, of Harvard's Solar Geoengineering Research Program (Bill Gates is a backer). Keith's group aims to release a plume of calcium carbonate particles—chalk dust, basically—some twelve miles up in the stratosphere, and then measure any resulting change in the reflectivity of the sky. He proposed this experiment in 2014 but it has still not been carried out. One reason is the fear that it will be a slippery slope to bigger field trials, normalization of the technology, and eventual deployment on a much larger scale. But, such is the potential importance of the method, to buy us time (as we've seen), we should do what we can to accelerate the trial, and launch more tests. If we don't do it responsibly, then the chances are someone will try it independently.

In 2010, US Congressman Brian Baird imagined just such a scenario. "Let us suppose," he told the House Committee on Science and Technology, "I am on the Maldive Islands, and I quite fairly and realistically assume the likelihood of the industrialized world actually cutting CO_2 emissions in a reasonable time is grim, and it is existential for us." What is to stop the Maldives trying to manipulate the climate directly to try and halt the warming, and save their islands? "What is to prevent that—or, even in a James Bond scenario, some rogue rich guy puts some airplanes in the air and seeds the clouds?"

Baird's James Bond villain has been called Greenfinger: a (possibly) well-meaning billionaire who attempts to engineer the climate for the greater good. Well, that's us in this chapter, but we'll try and do it as aboveboard as possible, with agreement from governments and ethics committees and collaboration with multiple research groups. What we don't want, as a climate scientist remarked to me, is a Russ George situation.

In July 2012, George, an American entrepreneur, sailed with eleven crew members to a swirl of currents called the Haida eddies roughly two hundred miles off the Canadian Pacific coast. There he started dumping more than one hundred tons of red-bronze iron dust into the ocean. The next month he returned and added a further 20 tons. His reasoning was that the iron would stimulate the mass growth of photosynthetic plankton, which would feed millions of salmon, thus helping the struggling village of Old Massett, which relied on the fish for its way of life. But the huge bloom of plankton would also, George hoped, capture many thousands of tons of carbon dioxide from the atmosphere and lock it safely at the bottom of the ocean when it died.

The plan didn't really work out. When news of his venture leaked, George was labeled an eco-terrorist and a rogue geoengineer; *The New Yorker* called him the world's first "geo-vigilante." Scientists have not been able to verify independently if his experiment worked or not, and many feel the venture was so irresponsible that it should not be repeated.[10] Well, it was indeed irresponsible, but I'd like to see some serious investigation into the idea. Assessing whether or not carbon dioxide from the atmosphere can be reliably sequestered to the depths of the ocean is a monstrously difficult task, but it's worth funding a proper research program to find out, just as David Keith's plans to make even a limited test of solar geoengineering should go ahead as soon as possible.

There is an organization, the Solar Radiation Management Governance Initiative, that oversees this sort of thing. They have allocated, for example, $430,000 to developing countries to assess how SRM might impact them. This is an area we can boost. Any kind of SRM should emphatically not be something decreed solely by Harvard, however well researched it is. In 2019, the US government gave $4 million to the National Oceanic and Atmospheric Administration to carry out research into solar radiation management.[11] We will empower researchers around the world, in South East Asia and in African countries, and include their findings in an overall decision.

∽

WHAT'S THE BEST-CASE SCENARIO? Putting aside the absolute best case, which is that the world makes huge emissions cuts and transitions to a carbon-neutral economy in the next ten to thirty years, our best-case scenario for SRM is that extensive trials show that its side effects are not too bad. That it doesn't wreck the monsoon in South Asia, for example. Or that the benefits of a temperature decrease are not offset by a reduction in the yield of crops, which was one conclusion of a 2018 report into SRM.[12] As we conduct field trials we'll start to understand the risks more clearly.

It's worth pointing out the scale of the problem, the difficulty in understanding what adding sulfate particles to the atmosphere will do. (Keith's test is with calcium carbonate, but the likely best substance for SRM is sulfate.) We've added hundreds of billions of tons of carbon dioxide to the atmosphere, but making precise predictions about how that will warm the planet is still beyond us. Adding a *couple pounds* or so of calcium carbonate, as in the limited initial Harvard experiment, is hardly going to give us a definitive answer.

But let's say we've performed a large number of trials, scaling up each time, and have garnered enough positive data and political and social support to draw up a manifesto of responsibility and pass an international treaty, and we decide to go for a global attempt. We'll need specially made aircraft that fly in the stratosphere and release their payloads of sulphate particles. Following research by Wake Smith at Yale and Gernot Wagner from New York University,[13] we'll commission a fleet of autonomous dronecraft with giant wingspans, capable of cruising in the stratosphere, steadily releasing their sulfur payload.

We'll need to purchase an island somewhere in the tropics, build a runway, and a port to receive shipments of sulfur. We'll allocate $6 billion for the setup. The running costs per year to sustain ten thousand or so flights will be in the order of around $300 million. Not much, but if we started it we couldn't stop. A sulfate shield only lasts a year or so: it drifts slowly back to earth (the effects of the "fallout" is another unknown). Not until we have pulled huge amounts of carbon out of the atmosphere can we let the shield come down for good.

It's been another of the great fears of SRM, that it commits us to indefinite use; if a shield suddenly came down, we would face what scientists call "termination shock," a shuddering increase in temperature, like when you step from an air-conditioned car out into the desert. As an aside, we would change the color of the sky. No one knows exactly what will happen, but putting particles in the stratosphere would scatter sunlight differently. The sky might not have a nightmarish Munchian tint, but it will probably be perceptibly whiter than now.

If, and again it's a huge if, the world decides that on balance the benefits of solar geoengineering outweigh the risks, we will still fully prepare for those risks. If we predict increased drought in some regions, or floods in others, as a result of the solar shield, then we should allocate money to help those areas. So, before any expenditure is signed off, we will fund global research, field trials, and outreach into sulfate injection, assigning $300 million to initiatives around the world. You can see we're hardly denting our pile of money: This work is vital, potentially world-saving and, initially at least, relatively cheap. We must get on with the research right away.

And there are a couple of very real dangers to be aware of. The first is that a flood of money into geoengineering research could tilt the public and politicians into pushing for deployment, say of a solar veil, before we know what we're doing. The second is that it could loosen even the slight brake on carbon emissions we're trying to make. Again: Emissions cuts are essential and unavoidable.

THERE ARE NUMEROUS OTHER PROPOSALS for different forms of geoengineering. A massive project, currently being examined, is the idea to stop the Thwaites Glacier in West Antarctica from collapsing. Thwaites is a brute, a glacier the size of Britain that is melting at a rate of 39 billion tons of water a year. Alone, Thwaites contains enough water to raise global sea levels by 26 inches. But, even worse, if it collapses, the rest of the West Antarctic ice sheet could follow, dooming us to a catastrophic 10.8 feet of sea level rise. The International Thwaites Glacier Collaboration has $50 million devoted to understanding the glacier; we could give it more to potentially prop it up, or stop it sliding into the ocean.[14] As usual, it is better to cut emissions than to have to attempt a project on such a scale—but let's look into it. Pumping water from under the glacier onto its surface might be an option.

There are other options for direct cooling, too. James Lovelock, the originator of the Gaia hypothesis, favors a space-based approach, where thousands of spacecraft deploy a sunshield in space to deflect sunlight from Earth. The advantage is that it could easily be collapsed if anything went wrong; the problem is it would cost tens of trillions of dollars.

So let's just look at one other method of geoengineering, the idea to brighten clouds.

In the 1990s, climatologist John Latham, then at the National Center for Atmospheric Research in Colorado (he's now at the University of Manchester, UK), started developing ideas for ways to reduce the amount of sunshine reaching the planet's surface in order to offset the global heating effect of climate change. Latham was fascinated by something called the Twomey effect, which shows that the amount of solar radiation reflected back into space depends on the concentration of droplets in clouds. Latham realized that you could increase this concentration in marine clouds by seeding clouds with tiny droplets of salt water.[15] We know from satellite images showing "ship tracks"—the equivalent to contrails left by airplanes—that clouds can be seeded by the sulfate emissions from ships, and Latham and colleagues have produced models showing how ocean stratocumulus clouds may be brightened by seeding with water droplets, and even how, if done in the Arctic, the cooling would start to restore sea ice.

On paper it all looks very promising, but testing it for real is quite another matter. That will require the development of some system able to deliver an ultra-fine spray of seawater into the lower atmosphere (the marine boundary layer) over a large area. Stephen Salter, an engineer at the University of Edinburgh, has well-advanced plans for this, having worked on proposals for remotely operated drone ships able to deliver the spray.[16]

I got in touch with Salter and, after we'd had a few back and forths, he sent me a photo of Neymar, the soccer star. Along with the attachment Salter had written simply: "Here is a picture of a happy young man." The photo showed Neymar in 2017, beaming as his transfer to Paris Saint-Germain was announced, which had cost the French club $263 million. I was perplexed, but Salter's point was made clear in his next email, which listed the costs of his cloud-seeding project. I realized that for the price of Neymar we could conduct all the preliminary trials, and then—if the trials suggest the method indeed does work as intended, and brightens clouds—for two years we could run an entire fleet of ships that could potentially start to restore the damage done to the Arctic since preindustrial times.

Again, this barely makes a dent in our budget. It also seems less risky than putting sulfates into the atmosphere, but there will probably be some residual effects on rainfall in other parts of the world. It is also attractive because it is a local form of geoengineering. At least at first, we would try it in the Arctic. Later we might modify clouds in the tropics.

Kelly Wanser is director of SilverLining, a geoengineering NGO based in Washington, DC. She says that, among ideas to prevent Arctic collapse, the most viable involve increasing the reflection of sunlight from the atmosphere. Wanser, who is also an advisor to the University of Washington Marine Cloud Brightening Project, bemoans the lack of investment in sunlight reflection. "This leaves us," she says, "with an enormous exposure to near-term climate risk and not enough fast-acting options to keep warming within safe levels."

We will fund immediate research into cloud brightening, to the tune of half a dozen Neymars.

⟨◦⟩

WHILE WE ARE INITIATING URGENT RESEARCH into solar geoengineering, we will accelerate attempts to remove carbon dioxide directly from the atmosphere.

One of the lesser-known details of international climate agreements, and of the IPCC scenarios for future warming, is that many of the predictions for temperature rises assume that by the second half of this century we will start removing large quantities of carbon dioxide from the atmosphere. This is known both as CDR—carbon dioxide removal—and NET, negative emissions technologies. The big problem is that we can't do this yet at any kind of scale to make a difference.

For all its world-changing power, carbon dioxide is a trace gas, making up just 0.04 percent of the atmosphere. That makes it very difficult to get out. We currently have ways to do it, but only on a small scale: We need to do it on a planetary scale, without adding to the carbon dioxide already emitted. Modelers and policy-makers assume that in the future this problem will have been solved, but it's not at all clear that it will be. One way would be to suck the stuff straight out of the air; another is to grow plants for biofuel, burn that to generate electricity, then capture the carbon dioxide produced and bury it underground. A third avenue to pursue is to plant trees. This has the happy side effect, if you do it properly, of increasing biodiversity, but the not-so-happy side effect of decreasing the amount of land available for other things we need, such as agriculture.

Some carbon-capture enthusiasts draw inspiration from the German Nobel prize–winning chemist Carl Bosch. Bosch took Fritz Haber's method for producing ammonia by pulling nitrogen from the air, and over a decade managed to reproduce it on an industrial scale, in what became known as the Haber–Bosch process. The ability to produce ammonia on this scale utterly changed the world, for good and ill, leading the science policy analyst Professor Vaclav Smil to name the process the most important invention of the twentieth century. If there was a similar breakthrough for capturing carbon dioxide it would surely qualify as the twenty-first century equivalent.

We are at a similar point to when Haber had figured out how to get small amounts of nitrogen from the air. But our job is tougher. For a start, we don't have a ready market for carbon dioxide, unlike the huge demand for fertilizer that the Haber–Bosch process fed. And nitrogen is 78 percent of our atmosphere. With carbon dioxide at such lower levels, you have to work much harder to get the stuff out.

That, however, is where we are, and it is a challenge being pursued around the world. The most advanced kind of carbon removal process is direct air capture (DAC): Pluck carbon dioxide straight from the air, and squirrel it away somewhere safe. It can be done, but it's expensive, with current costs at around $600 per ton of CO_2, though cost estimates of commercial-scale capture are from $30 to $300 per ton of CO_2. Sam Krevor, at Imperial College London, and colleagues, found that the prohibitive cost of carbon capture is what is limiting its greater use rather than any technical barrier.[17] A significant investment from us will smooth out the bumps to it being used much more widely.

If one person can be said to be the pioneer in direct air capture, it's Klaus Lackner, who has been working on the problem for at least twenty years. A physicist by trade, he has been refining mechanical-chemical methods to capture carbon dioxide and envisions making devices cheap enough that they can be mass-produced and used to capture billions of tons of the stuff, and, crucially, won't use so much energy that their own emissions become a problem. At the Center for Negative Carbon Emissions at Arizona State University, he has built prototype carbon-capture machines that drag CO_2 from the air onto dangling strips of treated plastic and form bicarbonate. The plastic strips are then folded into a water bath where the bicarbonate is converted into carbonate. When the water is drained, carbon dioxide is given off and captured.

Lackner wants to scale the process up, such that solar-powered devices capture and process the carbon and combine it with hydrogen extracted from water to make synthetic fuel. It's not cheap to develop, and it won't be cheap to run. The only way Lackner sees it working is if carbon is taxed.

These days he has a lot of competition. Climeworks is a Swiss firm that is trialing a number of carbon-capture projects, the most ambitious of which is in Iceland. There, carbon-capture units running on the country's geothermal energy collect 55 tons of carbon dioxide a year, pump it underground, where it reacts with basalt and turns to stone. But 55 tons per year is nothing. The Iceland plant is expanding to a projected 4,400 tons, but it's not until a later plant of 110,000 tons-capacity is running that we will start to see some economies of scale. At present, Climeworks charges its "pioneers" $600 per ton to capture "their" carbon dioxide. We need to get this cost to about $100 to make it viable.

Climeworks say they want to capture 1 percent of global CO_2 emissions by the mid-2020s, which would require them transforming from a small firm that builds its devices by hand to something like a major car manufacturer, building the machines autonomously. Carbon Engineering is another DAC firm, started by Harvard's David Keith, again with backing from Bill Gates, that is attempting to scale up the technology and bring down the price. They have partnered with fossil fuel companies, notably Occidental, to use their captured carbon dioxide (a little ironically) to recover oil in near-used-up wells. But their plans are ambitious and they are currently building a plant to capture 1 million tons of CO_2 each year. Their plan is to recycle the CO_2 into synthetic fuel, which would be genuinely carbon neutral.

Another carbon-capture firm, CarbonCure Technologies, has attracted investment from the likes of Amazon and Microsoft to capture CO_2 produced during the manufacture of cement—one of the most carbon-intensive of all industrial processes—and incorporate that CO_2 into the finished product.

We should invest in this technology, but there are caveats. These are proven ways to mop up excess CO_2, but it is hard to see how they can rid us of the hundreds of billions of tons of the stuff we've already emitted. We would need 9,500 of Carbon Engineering's 1-million-ton plans just to offset the world's two billion vehicles. And that's just addressing transport. DAC may play a vital role in fixing the emissions we can't avoid (from aviation, for example, and industrial processes such as cement and steel manufacture) but it shouldn't be allowed to distract us from going carbon zero as a matter of urgency.

Our money alone won't be enough, certainly. Let's take the current cost of carbon capture and sequestration, at $600 per ton. There are around 3.5 trillion tons of carbon dioxide in the atmosphere. If we spent our entire pot of money on capturing it with current technology, we'd capture only 1.9 billion tons of the stuff. One part per million of carbon dioxide in the atmosphere equates to around 2.35 billion tons of emissions. All our money wouldn't even change the dial. But if we can get the price down, we can and should use carbon scrubbers to offset those parts of the economy that take time to decarbonize.

So, we should fund the start-ups developing this technology, and stimulate the competition to improve it by creating incredibly lucrative competitions. We will give $100 million to the company that can create new sorbents that can fix and release carbon dioxide without the expensive heating process currently required, and that can fix a ton of CO_2 for $200. Additionally we will offer $100 million to the first company able to remove and safely bury 100,000 tons of CO_2 in a year.

⌒⌒

OUR ROLE IS TO KICK-START the carbon-capture business, to scale it up, and get it down in price, so that governments can take over. If DAC can deliver at a cost of less than $100 a ton, which is not impossible to imagine, then it may be that it can be financed through taxation.

Some useful provisions already exist. In 2018, the United States launched a tax scheme called 45Q that pays $50 per ton for companies to bury carbon dioxide underground. This sort of stimulation might help to force into existence a proper market for carbon. A more robust means is to charge a carbon tax when fossil fuels are burned and carbon dioxide is emitted. Australia did this in 2012, charging industrial emitters $24 per ton of carbon dioxide released. No one likes paying tax, and by mid-2014, the tax had led to a reduction in emissions of 17 million tons. But that same year, Australia (not now known for its climate leadership) scrapped the tax.

Sweden and Norway have had carbon taxes since 1991. Sleipner gas field in the Norwegian North Sea (it is named after the eight-legged steed of Odin) is the world's first offshore carbon-capture and storage plant. Natural gas trapped underground is not pure; it needs processing before it can be burned. In Sleipner, the gas is a mixture of methane and about 10 percent CO_2. Usually fossil fuel companies just separate the mixture and vent the CO_2 into the atmosphere, but in Norway the tax forces them to be more responsible. The government charges operators $50 per ton to emit CO_2 into the atmosphere, so the Sleipner owners, Equinor (formerly Statoil), scrub the CO_2 from the methane and compress it, then inject the fluid into sediment-filled aquifer about half a mile below the seabed. The aquifer is sealed by a layer of impermeable shale, which keeps the CO_2 safely trapped. Everyone wins: The government earns tax money and the fuel company saves paying it.

But the system obviously works only because there is a carbon tax. Statoil spent $80 million to build the capture and storage plant in 1996, but because they've been injecting 1 million tons of CO_2 per year into the sediment they save $50 million annually on tax. Their investment soon paid off. The aquifer is part of a formation estimated to be capable of storing 600 billion tons of CO_2, which is more than 100 times the total annual output of all the power stations in Europe. Many similar sites exist; the US Department of Energy estimates there is capacity of between

1 and 3.6 trillion tons in deep saline aquifers in the continental United States.

But nothing will be done until there is a proper economic incentive to do so. Dieter Helm, an economist at the University of Oxford, says that a carbon price is the only realistic way to do this. The European Union has the Emissions Trading Scheme, a permit-based scheme to try and arrive at a price for carbon. But Helm's preferred method is to set a proper price by using a carbon tax. In this way, very simply, goods and services that are more polluting in terms of carbon emissions—everything from hamburgers to airline flights, concrete construction to oil-fired power stations—cost more. Helm argues that, since the Paris Agreement is legally unenforceable, it isn't working and that a carbon tax is key to getting emissions down.[18] "A carbon tax makes it more expensive to pollute," he says. "We need R&D, innovation, and infrastructure but we do need a price on carbon."

The rules of Project Trillion prohibit direct political lobbying, but what we can do is commission carbon capture and storage (CCS) plants in suitable locations, near geological storage sites. We will start sequestering carbon dioxide as cheaply as we can. As we ramp up the scale we'll bring prices down. We will allocate $100 billion to this end, primarily by investing in existing CCS firms such as Climeworks and Carbon Engineering. If that cost sounds high, remember that Hurricane Katrina caused $80 billion in damages and Superstorm Sandy $65 billion. We'll invest another $100 billion in developing ways to run our CCS plants on renewable energy.

We will work with the fossil fuel industry on this, as they have the expertise and infrastructure in place to store carbon dioxide. Not to mention that they have some culpability for putting most of the stuff in the atmosphere in the first place. We can hope that economic self-interest will get carbon taxes onto the statute books around the world. Several governments, including the UK, have made net zero a legally binding target, so they will have to do something to make it happen, and they will need the money.

∽

WE WILL NEED A PORTFOLIO of methods of carbon removal. Two other compelling ideas—one land-based, the other in marine environments—aim to boost the natural ways that absorption of carbon dioxide process occurs.

Weathering is what happens when rocks wear down and are exposed to carbon dioxide, and it takes place over millennia both in the atmosphere and the ocean. Either way, the rocks slowly react with the gas and lock it up as an inert mineral. Enhanced Weathering is a new term for speeding up this natural process on the land, by grinding magnesium- or calcium-rich minerals into powder to facilitate their reaction with atmospheric carbon dioxide. This turns them into safe and inert magnesium carbonate or calcium carbonate. If the ground-up minerals are dumped into the ocean the accelerated weathering takes place there and is termed Ocean Alkalinity Enhancement (OAE). It has the huge additional benefit of countering the acidification of the oceans, which is a major threat to marine ecosystems, as we saw in chapter 4.

One of the prime candidates for OAE is a green volcanic mineral called olivine. As it comes into contact with carbon dioxide, olivine forms carbonate compounds that are locked up on the ocean floor or used in the shells and skeletons of marine life and *then* locked up on the ocean floor. Powdered or ground up olivine increases the surface area of the rock and speeds up the weathering and, if carried out on a large enough scale, it could have a huge impact.

One idea is to use olivine on beaches and in the ocean and let the action of the waves grind up the rocks. Roelof Schuiling and Poppe de Boer at Utrecht University in the Netherlands suggest that if olivine was spread over the shallow continental shelf around the coast of Britain, Ireland, and northern France, it would fix up to 5 percent of the world's carbon dioxide emissions.[19] That would be a massive project, requiring 0.08 cubic miles of olivine spread over 13,500 square miles of shallow ocean. We don't understand the effect this will have on marine life, and it might need some work to get the public on board with dumping millions of tons of rocks into the oceans, or turning our beaches green. But olivine looks promising and we'll invest in companies such as Project Vesta (a San Francisco nonprofit) in order to speed up research.

We will also invest in enhancing the process of land weathering. One major study in 2020 found that sprinkling half of the world's farmland with basalt dust—a mineral that turns carbon dioxide into carbonate—could fix 2 billion tons of CO_2 per year.[20] The lead scientist on that project is David Beerling, of the University of Sheffield. He says on our budget we could easily get large-scale deployment of enhanced weathering in the key locations of China, India, and Brazil. "You probably could do it for the bargain price of fifty to a hundred million dollars over a ten year time horizon," he says. "You could cost in massive expansion of enhanced rock weathering operations in the US corn belt for another fifty million, ditto for European countries." As well as drawing down carbon dioxide, this would help to restore critically degraded agricultural soil and will improve crop yields.

We'll allocate $200 million for Beerling's plan straightaway, and set aside $2 billion to roll out global initiatives for both kinds of enhanced weathering—assuming all the trials go well.

⌒

BIO-ENERGY CARBON CAPTURE AND STORAGE, or BECCS, is one of the oldest ideas for getting carbon dioxide out of the air. The concept here is that crops and trees are grown for burning in power stations, but their waste carbon dioxide is caught and buried. All projections for keeping temperature rise to 1.5° or 2°C (2.7° to 3.6°F) rely on CCS or BECCS being deployed on a truly stunning scale. People talk about fifteen thousand facilities on the scale of the Sleipner plant in Norway.[21] If that were all down to BECCS, it would mean devoting a land area two to three times the size of India to grow plants for burning. As we've seen in chapter 4, that land cannot be used without capitulating to the mass extinction we're already seeing, and adding to the deterioration of the Earth system. We're already losing too much land to livestock rearing and degradation from poor agricultural practice. But, if we could roll out BECCS on a grand scale on a sustainable basis, we could get as much as 5 billion tons of CO_2 per year out of the sky.[22]

We will need to start a verified carbon offset scheme. Pay to offset your emissions and we'll suck CO_2 out of the air. We'll offer services to manufacturers so they can sell products that are guaranteed carbon neutral or even carbon negative. Instead of wearing a designer label or being seen drinking Crístal Champagne, consumers will show off their negativity by buying offset products. People choose organic food now; we need to convince them to buy "C-neg" food. Microbreweries selling locally made beer have become an important part of the retail sector. It's hip to drink craft beer, but how much cooler would these companies be if they captured their emissions? If we got multinational companies involved, we could really start to make a difference: A new Coke Zero could be the coolest drink on the planet. Airlines could offer premium C-neg seats, with the excess cost being used to pay directly to capture and sequester carbon. It would allow large businesses to take a lead on climate change by demonstrating that they are willing to take a small hit.

The public response might bear them out. Some people might be willing to pay extra to live carbon-neutral lives. Dave Reay, a climate scientist at the University of Edinburgh, told me he is trying to balance his lifetime carbon debt by rewilding a patch of land he's purchased in Scotland. It will cost him many thousands of dollars, but it's a price he's happy to pay. We will be able to do the same, but with several orders of magnitude more money. The hope is that billionaires currently sitting on nest eggs for their children's future will decide that the best way they can realize that future is to start capturing carbon. They can still have their yachts, but it would be nice if they also win themselves a bit of climate respect by buying carbon. Or—for some of the more thoughtful—methane.

Most plans to capture greenhouse gases from the atmosphere center around carbon dioxide, the most widespread greenhouse gas, but the role of methane is not to be underestimated. Methane is far less widespread in our atmosphere than carbon dioxide; it has grown by 150 percent since industrial times but is only found at 2 parts per million. However, its effects are proportionately far greater on the climate. Over a century, it is twenty-five times more powerful but—crucially—it is eighty-four times more powerful in the first twenty years of its molecular life, after it is released. Removing methane thus has a big, fast impact on climate.

Pep Canadell and Rob Jackson of the Global Carbon Project, based at Stanford University, have proposed a technique to restore the concentration of methane to levels found before the Industrial Revolution. Their idea is to remove all methane in the atmosphere produced by human activities by oxidizing it to carbon dioxide. This seems counterintuitive: We end up with more carbon dioxide (turning 3.5 billion tons of methane into 9 billion tons of carbon dioxide). However, the CO_2 will disperse far slower and can be subsequently captured. The effect, according to Canadell and Jackson, would be to reduce global warming by one sixth.[23]

We will invest in the development of methane removal technology at scale, as well as other greenhouse gases, such as nitrous oxide.

∾

WE'VE BEEN LOOKING at what appear to be big numbers: aquifers that could store billions of tons of carbon dioxide, taxes that lead to emissions reductions of 19 million tons. Let's put that into perspective. In 2018 alone we emitted 41 *billion* tons of carbon dioxide.[24] What if—and bear with me if this suggestion sounds laughably simple—we just planted more trees? A team at ETH Zurich says that the planet can support a third more trees than it currently has, an amount that would draw down and store 226 billion tons of carbon dioxide. Wow! Given that each part per million of CO_2 is equal to 2.35 billion tons, that would bring the CO_2 parts per million down to about 320, which is what we had in the mid-1960s.

Is this not the best way to buy us more time? Of course, there are lots of details to work out. Simon Lewis, of the University of Leeds, thinks the 226 billion tons is an overestimate. And there are many demands on spare land—not least agriculture, housing, and recreation. But it does seem that there is a lot of disused and currently wasted land that we could redevelop in a massive tree-planting scheme.

The Zurich team, led by an ecologist named Tom Crowther, used data on forest cover from Google Earth and a machine-learning algorithm to predict which new areas could support forests. Once you take away areas where there are already farms and buildings, there is enough land around the world for nearly four million square miles of forest, an area about the size of the United States. Six countries account for most of the spare space, mostly because they are big and have already been deforested to a large extent: the US, Canada, Russia, China, Australia, and Brazil.[25] In this scheme, we'd plant around five hundred billion new trees.

How much would it cost? No one really knows. On a smaller scale, the government of Ireland says it will plant twenty-two million trees a year for twenty years, making a total of 440 million trees. The European Union is considering a proposal to plant 3 billion new trees by 2030.[26] This is the sort of policy we could support. In China, where a shocking 27 percent of the country is desert or has become desert, affecting four hundred million people, a "great green wall" is being planted in the northwest under the Three-North Shelter Forest Program.[27] Some 66 billion trees have been planted so far: We will certainly want to ask Chinese ecological engineers for advice. Another group we would approach for help (and with armfuls of cash) is the UN Convention to Combat Desertification.

In sub-Saharan Africa another Great Green Wall project is attempting to plant trees and vegetation across the entire width of the continent. So far, 68,000 square miles have been fenced off and planted or allowed to revegetate; the target is more than five hundred thousand square kilometers by 2030.

In a global initiative, many of the new trees will go on farmland, but that doesn't necessarily mean scrapping meat and dairy farming: In some areas trees can be planted at low density that doesn't interfere with grazing. In Kenya, the government has plans to plant two billion trees over two thousand square miles. But the costs haven't been worked out yet. Crowther said that some replanting projects cost only 30 cents per tree. If we could get to this kind of cost at large scale, we could plant one trillion trees for $300 billion. Obviously we need to persuade governments and landowners to allow us to use their land, and compensate them when necessary. Successful tree-planting schemes also need assurances that the trees will be tended and protected for decades. But our money would go a long way to getting this solution working. It seems most likely that a combination of eco-logically sensitive tree-planting and natural regrowth (as seen in chapter 4) will be the best solution.

The UN Framework Convention on Climate Change (UNF-CC) has done some work for us, agreeing to key activities needed to reduce greenhouse gas emissions related to forestry. As mentioned earlier, these are known by the acronym REDD+ (reducing emissions from avoided deforestation and degradation), and they commit to conserving and sustainably managing existing forest stocks and enhancing new ones. REDD+ is not straightforward (nothing is). Some forms of afforestation may mean converting grassland to plantation, which might mean a net loss of species richness.[28] But it can be done. In Ghana, a REDD+ program is helping generate roughly 650 million tons of emissions reductions by sustainably managing forests over 23,000 square miles.[29]

Tropical forests get all the attention, but we must not neglect other ecosystems, especially peatlands (bogs), drylands, and kelp forests. Peat bogs cover only 3 percent of the Earth but they are dense lumps of carbon. It's why peat has been cut and burned as fuel for centuries. Typically peatlands are 50 percent carbon and in total, around the world, they contain *twice* as much carbon as the forests. If they dry out, the carbon escapes, so it is essential to prevent this from happening wherever possible. Similarly, the UN Convention to Combat Desertification and the UN Food and Agriculture Organization say that restoring 3.5 million square miles of degraded land around the world could lock down enough carbon dioxide to get us some way toward the targets to stay below 2°C (3.6°F).

I've treated peatlands separately, but bogs, mangroves, and salt marshes are sometimes called "blue carbon." These regions are being rapidly lost, but protecting them could, according to a UN report, make up 10 percent of the emissions reductions necessary to stay at 2°C (3.6°F) of warming.[30] But it looks like we could do more than protect them. Kelp, especially, looks very promising.

Kelp is a giant seaweed, a macroalga, and grows incredibly fast—up to half a meter a day—and is often the dominant plant in coastal ecosystems. What we didn't appreciate until recently was that most of the carbon content of kelp drifts down to the seabed when the plant dies and becomes locked up in sediment. In this way an estimated 191 million tons of carbon are removed from the atmosphere each year.[31]

What we could do—and what we will do—is massively expand kelp forests around the world. The seaweed deacidifies the water—encouraging the growth of shellfish and promoting biodiversity—and can be turned into biogas and burned. If this was done on a big enough scale, and the CO_2 that was produced was captured, it could draw down many billions of tons of CO_2 a year, and, say proponents, sustainably produce large amounts of seafood to help feed a growing population.[32] Ocean afforestation could be a real lifesaver, perhaps at least on the scale of regular land afforestation. It is cheap, proven, scalable, and has multiple benefits.

The Climate Foundation at Woods Hole, Massachusetts, already has a plan for what it calls a marine permaculture system. You build a large frame about half a square mile, secure it eighty feet beneath the surface, and plant it with kelp. At this depth the farm doesn't interfere with any shipping passing overhead. Wave power is used to pipe cold water from the deep, bringing nutrients and promoting plankton growth. Repeat at scale, around the world, and we could start to sequester serious amounts of carbon—perhaps enough to buy us some time to change our addiction to fossil fuels. (It would do wonders for employment opportunities as well as global food security, too.) Kelp forests are declining as waters get too warm for them, so we urgently need to get on to this.[33]

Staying in the oceans, whale biologists have calculated that restoring whale populations to previous, preindustrial levels would help sequester massive amounts of carbon. Whales export carbon from shallow levels to the deep, as when they die their carcasses sink to the ocean floor. Rebuilding the blue whale populations would sequester four million tons of carbon in living biomass, comparable, the researchers say, to 166 square miles of forest. Restoring all baleen whales (of which blue whales are just one species) would store 9.6 million tons of carbon, equivalent to 425 square miles, an area the size of the Rocky Mountain National Park.[34] But more important than the carbon stored as living biomass in whales, the animals engineer entire ocean food webs. The presence of whales stimulates the growth of phytoplankton and the sequestration of carbon. The way whales modify the marine ecosystem could be harnessed to great effect.

A recent International Monetary Fund report proposes that if the population of whales returned to its pre-whaling size of 4 to 5 million (there are around 1.3 million today), this could lead to a significant increase in carbon capture.[35] The authors propose using a model similar to the REDD+ framework for preventing deforestation—in other words, by providing incentives to countries to promote whale protection.

SOLAR RADIATION MANAGEMENT, ocean alkalinity enhancement, and carbon-capture technology is what gets Silicon Valley excited and investors such as Bill Gates putting their hands in their pockets. Microsoft has announced a $1 billion project to offset all its emissions since 1975, mostly by funding a scale-up of direct air capture tech.[36] Jeff Bezos created and endowed the Bezos Earth Fund with $10 billion to find ways to "fight" climate change.[37] It seems that a more effective use of the money, certainly in the short term, would be to spend it on planting trees, restoring deserts, protecting and restoring wetlands and peatlands, sowing giant kelp forests, and increasing the number of whales.

We will need carbon capture to step up, however. Forests burn in wildfires, they die if they're not looked after or the wrong species is planted, and it's hard to quantify just how much carbon they're assimilating. And some of the land we might grow extra forests on will be needed for agriculture. Plus, as we've seen, some industries will be hard to decarbonize and we'll need to capture the polluting carbon. All that said, a mass program of tree planting and managed forest regrowth in key regions might be the single greatest investment we could make to protect the planet. I could live with strolling through the Bill & Melinda Gates National Forest or the Amazon.com World Wide Wood, the Elon Musk Giga-Forest or the GoogleForestPlex. I'd love to swim through thriving kelp forests knowing that they are helping us buy time to save the planet. What a legacy for billionaire philanthropists, and what a public-relations opportunity for trillion-dollar businesses.

We started this chapter with art inspired by natural geoengineering—and so shall we close. When Mount Tambora in Indonesia erupted in 1815, the ash cloud caused an abrupt change in the world's climate, and the "year without a summer" in Europe. Some scientists predicted the end of the world. Moved to prophesize apocalypse and the extinction of the human race, Lord Byron wrote "Darkness," which ends with these words:

The winds were wither'd in the stagnant air,
And the clouds perish'd; Darkness had no need
Of aid from them—She was the Universe.

We are at the point where the road diverges. We can take the path that averts the apocalypse. It's clearly signposted. We wouldn't even need to spend that much money.

Achieved

A greener planet, literally: one with vastly more photosynthesizing biomass. We have developed the means to capture carbon emissions so that hard-to-decarbonize industries have the time to get to net zero. We have an insurance policy: The understanding of what we'd be doing if the worst happens and we are forced to undertake planet-wide climate engineering.

Money spent

Field trials of solar radiation management:	$6 billion
Field trials of enhanced weathering:	$2 billion
Field trials of cloud whitening:	$1 billion
Restoration and protection of global peatlands:	$10 billion
Global tree planting and forest regrowth program:	$300 billion
Global kelp afforestation program:	$100 billion
Ocean engineering through whale populations:	$100 billion
Development of carbon capture at scale, powered by renewable energy:	$100 billion
Development of other greenhouse gas capture technology:	$10 billion
Commissioning of carbon-capture and storage plants:	$100 billion
Competitions: > to fix and bury 100,000 tons of CO_2	$100 million
> for a process able to fix 1 ton of CO_2 for $200	$100 million
Total:	**$729.2 billion**

The long-extinct auroch, the ancestor of modern cattle, painted on the cave walls at Lascaux around 17,000 years ago.

Turn the World Vegan

AIM: To transform global farming and initiate a genuinely green revolution that will slash the amount of greenhouse gases produced by agriculture. To intensify sustainable farming practices in order to avoid mass famine by the end of the century. To incentivize farmers and the public to change food production and consumption such that the use of animal products is minimized.

SO MANY OF THE ISSUES TACKLED in this book are interrelated: poverty and global health; biodiversity and climate change. So too with the food we eat. Our system of agriculture is shockingly polluting. In the UK, agriculture accounts for 70 percent of the country's land use but only about 0.7 percent of GDP, and its measured carbon emissions are 11 percent of the country's total. In the US, roughly half of all land is used for agriculture, which makes up 0.6 percent of GDP and 10 percent of emissions.[1] "Farming as a proportion of GDP is the most polluting industry we've got—much more polluting than oil companies," says Dieter

Helm of the University of Oxford. We can't hope to tackle the climate crisis without changing the way we farm and the way we eat. That's what this chapter is about.

But before we start working out how to buy the farm and ensure we can feed the world through sustainable agriculture, let's take a moment to consider how we got to where we are today. Or, more precisely, how one species changed and shaped our world. This is the story of the auroch.

∽

AROUND TWELVE THOUSAND YEARS AGO, the auroch, an imposing animal that lived across Europe, Asia, and North Africa, went about its business, which was grazing. It was a species of wild cattle and seems to have been revered by early humans, given the paintings on the cave walls of Chauvet and Lascaux. But ten thousand or so years ago it had the terrible misfortune to be domesticated, and selectively bred for milk and meat. Wild aurochs had long, relatively slender legs, making them almost as tall at the shoulder as they were long in the trunk, and were equipped with fearsome horns and small, hardly noticeable udders. Over the years, all this changed as artificial selection did its work. The udders of dairy cows grew, and the cattle became stubbier and bulkier, carrying far more muscle. As the domestic animal developed, the auroch itself was hunted to extinction. The last one died in the seventeenth century. But its descendants took over the world.

The Europeans took cattle with them to the Americas as part of the Columbian Exchange, the globalization of food species that began when trade was established between Europe and the Americas. Corn (maize), beans, potatoes, and tomatoes went east to Europe, while domestic cattle and chickens (and smallpox), and wheat and sugar, went the other way. And that's when the story of cattle goes up a notch. Cattle ranching and cowboys spread across the north and south of the continent, baking itself forever into the culture of the new United States. Eventually cattle farming became industrialized, until now, when there are more than a *billion* cows alive, according to the World Cattle Inventory, with 212 million of them living in Brazil alone. The business of raising them takes up 83 percent of global farmland, and produces 60 percent of the greenhouse gas emissions that come from agriculture. In terms of land clearance, they are the single greatest destroyer of the planet's ecosystems.[2] They must be stopped. Which is to say, *we* must stop farming them, certainly at such scale. Even if we could click our fingers and stop all greenhouse gas emissions from the burning of fossil fuels today, the emissions from agriculture would take us past 1.5°C (2.7°F) of warming, missing the goal of the 2015 Paris climate meeting, by the mid 2050s. We'd go past 2°C (3.6°F) by the end of the century.[3] "Agriculture is the single biggest threat to biodiversity and extinction," says David Tilman, an ecologist at the University of Minnesota.

We also have to revolutionize how we make food, if we want to feed the world. Today, of a global population of 7.7 billion, there are about 2 billion people who are overweight and obese, and 2 billion with nutritional deficiencies. There are some 800 million who are chronically hungry because of poverty. The population will go up to around 9.8 billion by the middle of the century, and food production will have to increase by 60 to 110 percent to feed everyone.[4] And, as poor and middle-income countries get richer, their citizens will want more meat. The average person in the United States eats 265 pounds of meat per year, compared to 9 pounds per person in India, and 37 pounds in Kenya.[5] Global demand for meat and animal products is predicted to rise by nearly 70 percent over the next thirty years.

The World Resources Institute commissioned an in-depth report into how to meet this demand: *Creating a Sustainable Food Future: A Menu of Solutions to Feed Nearly 10 Billion People by*

2050.[6] This describes the way to feed the world without further damaging ecosystems or increasing poverty while at the same time reducing greenhouse gas emissions.

Perhaps the single most important recommendation is to farm fewer cattle. Far fewer. We don't have to let cows go the way of the aurochs (though, if young cows could comprehend the lives ahead of them, they would probably welcome extinction), but we must massively reduce their number. To do that, we have to get 2 billion people to cut their consumption of beef by about half. Lamb, too. One way, and we have to push for this, is to make the price of meat reflect the environmental cost of producing it. The OECD says that agriculture gets $620 billion a year in subsidies, of which two-thirds is via measures which "strongly distort farm business decisions."[7] This kind of distortion leads to beef being half the price it ought to be in the United States. We'll leave it to the politicians to work out how to change that, but one thing we can do is invest in alternatives—plant-based food that mimics meat so precisely as to win over carnivores.

In this chapter we'll look first at the scale of the task of feeding the world. It falls into four parts: livestock production; unsustainable agriculture; the problem of waste food; and the cost to human health. We'll consider funding solutions for each of these problems, including an innovative way to change (and improve) people's diets.

❧

MEAT IS NOT FIT FOR PURPOSE. It's not quite as catchy a phrase as "Meat is murder," but it's how things are. Meat has negative health effects because we're eating too much of it; it is a significant contributor to climate change; producing meat depletes natural resources; and the animal welfare issues it raises are enough to turn anyone's stomach.

In the last chapter we met Johan Rockström, of the Potsdam Institute for Climate Impact Research. Rockström and a group of scientists have put together a blueprint for what they call a "safe operating space for humanity," a space bounded by limits that, if we cross them, will impact human health and destabilize the biosphere. The boundaries include things such as the rate of extinction (which we've already crossed), the nitrogen and phosphorus cycles (likewise), land use (also crossed), and climate change (at 415 parts per million of carbon dioxide, we're halfway across). The method of agriculture we have pursued for the last fifty or so years has contributed in varying amounts to the breaking of all these boundaries. Such is our impact on the planet, and such is our exploitation of it, that it's been said that the living world is best referred to, not as our biosphere, but as a global production ecosystem.[8]

Livestock production is unbelievably profligate. We use a quarter of the ice-free land on the planet to rear animals to eat. That's an area the size of Africa. According to the UN Food and Agriculture Organization, livestock farming generates 14.5 percent of total greenhouse gas emissions.[9] That's about the same as trains, cars, ships, and planes put together. Ruminants (for our purposes, cows and sheep) belch vast amounts of methane, which is why the impact of raising livestock is so high. If "cows" were a country, it would be the third-largest emitter of greenhouse gases after China and the United States.

Livestock animals also suffer on an industrial scale. Worldwide, more than sixty billion chickens are killed annually, the vast majority raised in conditions of appalling cruelty. Each year, tens of millions of breeding pigs spend their entire lives in crates that don't allow them even to turn around. Hundreds of millions of cows live in CAFOs (concentrated animal feeding operations), which are industrialized mass rearing sheds for animals that the Centers for Disease Control say can produce more waste per year than an American city.[10] Waste that runs off into the water supply. Intensive farming like this requires liberal use of antibiotics and other pharmaceuticals to prevent the spread of infection. This too leaks into the wider environment, driving problems such as the evolution of antimicrobial resistant "superbugs."

Many of the large mammals raised in intensive conditions are killed in gas chambers. They are suffocated, ironically, by carbon dioxide. Others are transported by ship from Brazil to be killed in Asia. What it's like for an animal to travel in one of these ships doesn't bear thinking about. There are tens of thousands of CAFOs in the United States alone, and many more around the world. They are quite simply an abomination, and it is unconscionable that we subsidize cruelty on this scale.

Putting welfare questions aside (with some difficulty), eating high up the food chain is simply a poor decision from a systems engineering point of view. The "plant to animal conversion efficiency" is a measure of the amount of plant matter you have to feed to an animal to get flesh or product out at the end. For chicken the conversion efficiency is about 12 percent, for pork 10 percent, and for beef 3 percent. Milk is 40 percent and eggs 22 percent. The appalling inefficiency of plant-to-beef conversion means that a third of the global production of cereal is fed to animals, and it's why it takes 1,800 gallons of water to produce a pound of beef.[11] Hundreds of thousands of tons of soybeans are fed to cows every day in the Amazon.

All this, of course, means that there are big savings to be made if we stop eating so much meat. If Americans didn't eat beef, 42 percent of crop land in the country currently used to grow animal fodder would become available. The point is that people need not give up milk or cheese or even pork or chicken to have this benefit. Beef production is so inefficient, and so planet-heating, that dropping it alone has a huge impact.[12]

This is crucial when it comes to the message we need to get out to people: You can achieve a lot by making a small change. "Reducetarianism" works. Meat-free Mondays. VB6—vegan before 6 PM—each day. Reducing the amount of meat you eat is a great start. After I tried giving up animal products for a month for "Veganuary" I found the tendency to choose vegan products carried on (I almost said "bled over," but it doesn't seem right) into the rest of the year. Oat milk, vegan wine, vegan cooking and eating and consuming in general—it's becoming much easier to enjoy food without animal products, and our job is to make it even more so.

Comparing different methods of producing foods reinforces the profligacy of meat. Joseph Poore at the University of Oxford, and Thomas Nemecek at Agroscope, an agricultural research institute in Zurich, looked at a mass of data from 38,700 farms selected from the world total of 570 million. As might be expected with such a varied data set—farms have different ways of doing things and different costs and benefits—there was a large range in the environmental impact of making the same product in different farms, but the lowest-impact animal products almost always exceeded the impact of vegetable substitutes. In other words, the most emission-lite form of making some animal product is still much worse than the most profligate vegetable. For example, the most "environmentally friendly" beef has thirty-six times greater greenhouse gas emissions than peas, as well as six times greater land use.[13]

Animal products fail on all environmental and moral measures. Sure, you can justify meat on grounds of culture. Of course. No one ever celebrated a victory or a wedding with a feast of carrots. I understand, and even applaud, Anthony Bourdain's boundary-pushing spirit of adventure ("Vegetarians are the enemy of everything good and decent in the human spirit, an affront to all I stand for, the pure enjoyment of food") and I'm acutely aware of being a middle-aged white man presuming to wag his finger. But enough is enough. We have to quit the industrial farming of animals and reimagine how we can cook without them—at least until replica meat is up to scratch.*

⌒

NONE OF THIS IS TO SAY that crop production is a problem-free enterprise. We currently use 1.5 billion hectares of the ice-free surface of the land for crops, with another half a billion hectares that has been cleared, used, exhausted and abandoned in the last fifty years. Hundreds of millions more hectares will be needed in the next forty years to meet demand, and most of this, if we don't do anything about it, will come from the tropics.[14] We can't afford this for biodiversity reasons, but mainly because the most critically important region, the Amazon rainforest, has already been deforested close to a tipping point of destruction, as we have seen.

The twentieth century's Green Revolution transformed agriculture across the world. Its central figure was American agronomist Norman Borlaug, who pioneered the use of high-yield varieties of crops and combined them with intensive agricultural methods in India, Pakistan, Mexico, and other countries. He won numerous awards, including the Nobel Peace Prize, but the most extraordinary accolade associated with Borlaug is that he is credited with saving over one billion people from starvation. This was an unparalleled achievement, but the environmental cost has been huge—from the massive use of fertilizer and pesticide to bring about the yield increase, and from for the soil erosion and insatiable thirst for water that goes with intensive agriculture. Borlaug's methods also shifted the way the Earth system operates. Just four crops—wheat, rice, corn, and soybean—provide more than half the world's calories. (There are more than two million varieties

* The food of South India is a good place to start.

in seed vaults.) These are grown in vast monocultures, bathed in synthetic pesticide and nitrogen fertilizer to get spectacular yields.

Industrial-scale fertilizer producer Carl Bosch (whom we met in the previous chapter) is the other hero-villain here. Farmers use 127 million tons of chemical fertilizer a year, a large percentage of which goes into the atmosphere as nitrous oxide (which, like methane, is a more potent greenhouse gas than carbon dioxide) and leaks into the water supply as nitrates and phosphates. Our soils are stripped of the carbon they store. Intensive farming relies on over-fertilization, and regular tilling, which breaks up the natural structure of the soil and kills many of the microbes that live there. Both are factors in global decline in soil fertility and soil erosion, and loss of soil carbon.[15] It's an underappreciated fact that there is three times as much carbon in soils as there is in the atmosphere.[16]

In freshwater and ocean ecosystems this inadvertent over-fertilization results in massive blooms of algae, which subsequently die and consume oxygen as they decompose, causing sprawling seasonal dead zones around the world. Some 95,000 square miles of fresh water and ocean is so classified.[17] Another source of greenhouse gases comes when land is tilled and carbon dioxide in soil is released into the atmosphere. Both these forms of pollution are currently looking to double in the next thirty-odd years.[18] More water is used in farming than anything else we do and, as we've seen in chapter 4, more biodiversity is lost as a result of clearance for agriculture than anything else. It's why Rockström's planetary boundaries are busted or straining.

We're at Peak Farm. It's time to accept it. Stop clearing land for yet more farms, cling to the untouched land we've got left, protect it and nurture it, and in many cases restore it. Use a wider range of crops that have a better ecological fit for their environment and give a better yield. If they aren't good enough, or will struggle with climate change, use genetic modification to make them better.

WE EARLIER ELEVATED "COWS" to nationhood and found they rank third in the table of greenhouse gas emitters. If "waste food" was a country, it too would be in third place, contributing 8 percent of annual emissions. That's because 30 to 40 percent of all the food we produce is wasted.[19]

The waste happens for different reasons in rich and other countries, and at different stages of the consumption line. In developing nations it happens at the farmer's end: Crops are not fully harvested, not properly stored, and the infrastructure isn't good enough to get the food to the consumer. Storage wastage may be due to lack of refrigeration, which accounts for most of the wasted food in India,[20] or due to pests and microbial attack, which can account for a third of the rice crop in South East Asia.[21]

The use of "liquid air"—where renewable energy is used to compress air into liquid—is a novel technology that needs investment to break out. It is being developed primarily at the University of Birmingham Centre for Cryogenic Energy Storage. We saw in chapter 3 how hydrogen can be made through electrolysis as a way to store excess renewable energy, and it turns out that liquid air can perform a similar function, and as a low-emission refrigeration unit.[22] Collaboration between the Birmingham Cryogenic Energy Storage center and India's National Centre for Cold-chain Development will help develop low-carbon refrigeration in the subcontinent (and elsewhere) and start to reduce the problem of perishable food, cut food waste, and reduce emissions. We'll support a major program.

In rich countries, waste occurs at the consumer and retailer end. Consumers don't like wonky carrots, retailers will upsell with "three for two" offers and the like, or supersize portions, and "best before" labeling is confusing—all of which leads consumers to chuck away perfectly good food. If the proportion of wasted food could be halved, then 300 million hectares less land would be needed by 2050, with the associated savings in greenhouse emissions that would also bring.[23]

One way to get these savings is to recycle the wasted food. Not compost it but recycle it, by feeding it to insect larvae. Insects such as the black soldier fly have larvae that pretty much eat anything, growing into fat little maggots that can be dried and used as food—for pets, for fish in fish farms, and for us, to add protein to bread and biscuits, and to provide the fat in ice cream. As we've seen, 80 percent of the soy grown in the tropics is fed to animals, who are very poor at converting it to protein. Instead feed them dried maggots that have been raised on waste food. The grubs are 40 percent protein and 30 percent fat. A quarter of all fish caught from the dwindling stocks in our oceans goes to feed bigger fish in aquaculture.[24] Feed them insect larvae instead!

A comparison of insect sources of fodder and those from animals found that insect farming was more sustainable[25] and produces less greenhouse gas than composting waste food.[26] European and US regulators are getting on board with the idea: in Europe, insect protein is now permitted in fish food, and will soon be allowed for chickens and pigs; and the US already permits insects to be fed to chickens. There are several companies around the world that are developing insect farming. We will back them strongly and look to scale up as soon as possible.

◁～▷

PEOPLE IN THE RICH COUNTRIES of the world eat too much meat, sugar, and processed foods.[27] Food manufacturers have captured our evolutionary history and led us to like foods that are bad for us, because in the past those tastes were always good for us. Tangy and sweet used to signify nutritious fruit. Now it is refined starches and sugars. And we are hungry for blood. Americans eat

twice the recommended amount of protein, the majority coming from beef and chicken.[28] But it's similar all over the world. A
World Resources Institute study found that the average person
in over 90 percent of countries around the world eats one third
more protein than is recommended.[29]

The first analysis of the link between human health and our
diet and its impact on the environment was performed by a team
from the University of Oxford in 2016. They found that replacing
some of the meat in our diet with plant-based foods, following
WHO guidelines, could reduce the number of deaths around the
world by 6 to 10 percent. At the same time, this change in diet
would bring down greenhouse gas emissions attributed to food
by between 29 and 70 percent.[30] The financial benefit from health
and environmental savings was estimated at between $1 trillion
and $31 trillion by 2050. A follow-up study in *The Lancet* in 2018
found that, in rich countries, a switch to a vegan diet would lower
premature mortality by 12 percent and reduce greenhouse gas
emissions by up to 84 percent.[31]

But how do we lower the environmental impact of the meat
industry? One way is to reduce the emissions at source—in other
words, make production more efficient—and the other is to get
people to eat less meat.

In Australia the greenhouse gas contribution of livestock
is almost 10 percent of the country's total emissions, so much
work has gone into finding a way to reduce the amount of
methane that the animals produce. It turns out that seaweed is the perfect additive. Lab tests show that adding small
amounts of kelp to food prevents methane formation,[32] and
tests in animals showed reductions of 50–80 percent over a
seventy-two-day testing period.[33] This is good, and should
certainly be investigated and scaled up if possible. But it is by
no means the solution on its own. Unfortunately, other ways
to reduce emissions come at a welfare cost. Intensive farming

in vast sheds may have a lower environmental impact, but the animal suffering is off the scale.

The better way is to replace animal products altogether, using plants. Soybean plants produce twice as much protein for every third of a square mile as other vegetables, 5 to 10 times more than dairy animals raised on the same amount of land, and 15 times more than land used for meat production. The Climate Focus think tank, based in the Netherlands, found that 10,300 square miles of forest are cleared each year to graze cattle and to grow crops to feed livestock. Soy plantations account for 2,300 square miles of forest clearances each year. Simply put, we need to use the land currently given over to cattle, to raise crops for people. An investor group, the Farm Animal Investment Risk & Return (FAIRR) initiative, recommends that meat should be taxed just as other damaging products are, such as tobacco and sugar.[34] A meat tax would reduce consumption, improve health, and help provide a financial incentive for switching to crop production.

Soy is the basis of the famous Impossible Burger, the one that oozes "blood" and fat when you bite into it, just like the juices of a rare-cooked steak. The burgers are entirely vegan. The juices are from a compound called leghemoglobin, an iron-carrying molecule similar to the hemoglobin you find in blood. In July 2018, the US Food and Drug Administration, which regulates new food products, signed off on the safety of the Impossible Burger. I've not yet eaten one, because the leghemoglobin comes from genetically modified yeast, and in Europe we are funny about this stuff. I have, however, eaten the Beyond Burger, also vegan, which uses pea protein to mimic ground beef. I thought it easily the best burger I'd had in thirty years, but then it's been three decades since I ate meat, so what do I know? But my carnivorous friends seem to agree it is very close to the taste of real meat.

Meat is so mouthwatering because of a process occurring during cooking called the Maillard reaction. This takes place when carbohydrates and amino acids are fused in a hot, slightly moist environment, and produces the delicious aromas we associate with a barbecue or a grill. Most of what we think of as meat's taste is actually its aroma, and with careful cooking we can replicate it. And most of our craving for meat is actually a craving for fat, and we can replicate that, too. Blending small amounts of meat with soy protein can make hybrid burgers almost indistinct from regular ones. This also has a big effect on the carbon footprint of the product.

But it may take more than plant-based fake meat to get most people to stop eating animals. That's where lab-grown meat comes in. Actual animal flesh, but grown in a cell culture. This is currently prohibitively expensive: The first burger grown from animal cells in a lab, and eaten in a PR stunt in 2013, cost $286,000. But if the cells can be grown at scale, then we can start to offer "real" meat made without animal suffering, and business will take over. There is still some way to go. We need to grow the full range of cells that you'd find in a real piece of meat, not just the muscle cells. To make the product taste like an animal, it needs fat.

We also need to replicate other animals. A range of start-ups such as Impossible Foods, Memphis Meats, Just and Finless Foods in the US, SuperMeat in Israel, and Mosa Meat in the Netherlands, are betting that meat farming and even fish farming will be phased out. Pat Brown, CEO and founder of Impossible Foods, says he wants to bring an end to animal agriculture and deep-sea fishing by 2035. That seems like quite a tight deadline, but we will happily invest in it.

⌇

JOSEPH POORE, THE MAN BEHIND that study of the environmental impact of different food types across nearly forty thousand farms, says we also need to invest in ethical and environmental education, globally. Of course, as we saw in chapter 1, that will mean investing in basic education, too, in many countries. Poore thinks that education about ethics will itself drive a change in diet. Anyone who has thought deeply about the issues, he says, will realize the consumption of animal products, when animals have been exploited or the environment unnecessarily damaged, is unacceptable.

Simply making plant-based products and meat alternatives visibly available does have an impact. As an example, on a small scale, the University of Cambridge removed beef and lamb from all its catering outlets, and saw a 15 percent reduction in carbon emissions per pound of food purchased, and a 13 percent reduction in land use per pound of food purchased.[35] As plant-based and vegan food is offered in more sites such as schools and hospitals, more people will start to shift their consumption. But it is not enough. As we saw, there is a very tight link between rising income in poor and mid-income countries and meat consumption. We're never going to change how people eat by shaming them. Nor are we ever going to legislate what people can and can't eat—the best we'll be able to do is to tax meat.

Ecologist David Tilman has another idea: to run lucrative annual food competitions. Let's take just $1 million of our budget and set up an international competition of twenty prizes worth $50,000 each, for the best sustainable recipes. People send their recipes and photographs of their food to a website and vote on the best or a panel of judges selects the best. The point is that, under the rules of the competition, the food must meet sustainability standards and health standards. We set up a points system for the impact on water use, greenhouse gas emission, and so on. People will be able to use a small amount of beef if they want for flavor, but will have to come in under our strict environmental impact standards.

The most delicious food that meets these criteria wins. The beauty of it is that we use the creativity of the world to solve the problem. We don't try and impose restrictions on other countries or their cultures. We could set up a restaurant chain that sells the winning recipes, from Ethiopia, Nigeria, Sri Lanka . . . Tilman wants to call the competition and the restaurant chain *The Winners*, but we will be happy if, once we've started the investment, someone else wants to take over and put their name to it. The restaurateur and entrepreneur Kimbal Musk (brother of Elon), perhaps?

I'm biased, but what's not to love in these proposals? If we can make it tasty enough, and scale up meat-substitute production, we could replace a large chunk of livestock farming. Why wouldn't you eat the tastier and healthier and planet-saving option if given a choice? Or even, hopefully, the cheaper option, if we got the subsidies right. We could take a huge amount of pressure off the environment and reduce untold animal suffering.

～

WE NEED TO MODERNIZE agricultural methods with a view to increasing the amount of land that is intensively but sustainably farmed and to closing the "yield gap."

Modernization means improving fertilization, especially, but also irrigation. This includes the greater use of "climate-smart crops"—varieties that are bred for optimal growth in the local conditions, and also genetically modified crops that deliver greater yields, produce more nutritious food, or are more resistant to saline or drought or heat or some other adverse condition.

This needs to be done at all levels, from industrial mega-farms down to small-scale farmers. The key target is to close the gap in the yield obtained by smaller, poorer farmers and that achieved in the big farms. In the least developed countries, yields are 20 to 25 percent of what they could be if intensively farmed, and even in many developed countries yields are less than half what they could be. But of course the yield gap can't be closed by swamping the land with fertilizer and pesticide. It must be closed sustainably.

This is called sustainable intensification. In a nutshell it is about timing the application of fertilizer. Standard practice has been to dump all the fertilizer at the start of the growing season, but this leads to huge losses. Large amounts of nitrogen are simply washed away by rain and irrigation, or it turns into ammonia gas and disappears into the air. Timing the application of smaller amounts of fertilizer throughout the season can deliver high yields with a third less nitrogen. The UN Environment Programme found that a 20 percent improvement in nitrogen application would save 22 million tons of fertilizer and between $50 and $500 billion.[36]

In the US, a web-based tool called Adapt-N is being trialed to reduce fertilizer use. Farmers just need to access the tool on their smartphone, and based on soil and weather and crop information it tells them when to apply fertilizer and how much to use to get optimal growth. In a trial across 113 farms in Iowa and New York that compared conventional methods of applying nitrogen with Adapt-N, the tool resulted in lower use of nitrogen and no change in yield, and delivered 65 percent higher profits.[37]

Sustainable agriculture is starting to take off. A review of a massive data set of farming initiatives around the world found that 163 million farms—that's almost a third of the world total—have passed a threshold where they are classed as practicing some form of sustainable intensification. That accounts for 453 million hectares of agricultural land, some 9 percent of the world total.[38] We need to get behind this rolling change to the farming system and give it a push, and cease relying on synthetic chemicals.

Jules Pretty, of the University of Essex, who led this research, says that the creation of agricultural knowledge economies, made by building up social capital, is key to transformation. Farming is an endeavor that is shared by hundreds of millions of people. Connecting them, establishing and strengthening the bonds between them, increases trust and facilitates the sharing of knowledge. This is what Pretty means by social capital. We need to create innovation platforms around the world, and employ facilitators to build the social infrastructure and create the links between farmers and groups. Pretty is working on a second global assessment of agriculture, and he estimates that in the last fifteen years 8.3 million social groups have been formed across the agricultural community. He says we'll need 30 million such groups to connect all the world's farmers. With one facilitator to train six groups, and paying $20,000 to each as a salary, that means we should allocate $100 million from our budget for this.

Pretty makes the point that, while governments have passed regulations to ban certain harmful practices, or set limits for pesticide use, they are far less inclined to regulate to ensure positive practices. This is why our investment is vital. These social groups are what Paul Hawken of Project Drawdown calls the immune system for the world.

CHINA HAS ADOPTED SUSTAINABLE FARMING on a huge scale. Between 2005 and 2015 the country pursued an extraordinary program of sustainable agricultural change taking in 20.9 million farmers with a total of almost 40 million hectares of land. Thousands of researchers and agricultural personnel worked with smallholders on lands ranging from icy cold to subtropical, changing the farmers' behavior not simply with a demonstration of the science but by involving the farmers in the campaigns, increasing problem-solving skills, fostering cooperatives, building trust—as Pretty said, by developing social capital. The program taught a system of integrated soil–crop system management (ISSM) that helps determine the optimal crop that should be grown according to the average sunshine and temperature conditions, and the soil type, and the resultant needs of the crop. Average yields of rice, corn, and wheat increased by more than 10 percent, while nitrogen fertilizer use went down by 16 percent. This yield increase and cost savings on fertilizer were equivalent to $12.2 billion.[39]

This kind of ISSM system must be rolled out across the world. There are 2.5 billion smallholders globally, farming 60 percent of the world's arable land. Our current systems breed and grow crops and livestock that are inherently weak, and require chemical help to achieve high yields. We need to throw off the yoke of agrochemicals and (for livestock) pharmaceuticals, and develop with all means necessary a diverse range of foodstuffs that don't break or bend planetary boundaries. "The point about sustainable intensification is that it doesn't cost anything," says Lynn Dicks, an agro-ecologist at the University of East Anglia. "You avoid expensive pesticides and fertilizers and focus on maximizing nature's services, without a yield penalty." Dicks advises that we build a global network of agronomists and farmers, linking them so they can share best practice. Minimizing chemical use, maximizing ecosystem services, and breeding and selecting for farming those crops and livestock that complement the system of sustainable intensification. It's a bit like the global network of health care workers recommended in chapter 2.

We need to invest in agro-ecological colleges to train farmers, prioritizing lower-income countries and regions that are earlier on the intensification trajectory, such as Eastern Europe, parts of Africa, Latin America, and Asia. "We need an agro-ecological revolution and it has to start on the ground," Dicks says.

We saw how millions of hectares of crop land lie abandoned. Maybe over-irrigation has made them too salty, or the groundwater has been depleted, or the soil has been exhausted. This can be restored and regenerated. Soil carbon sequestration was an initiative announced at the Paris climate talks in 2015, with the aim of improving the carbon concentration in soils by 0.4 percent per year. Rattan Lal, of Ohio State University, found that soils have the potential to sequester 2.7 billion tons of carbon per year.[40] This will take some money but is going to be essential. The use of cover crops, mulches, and compost help to increase the nutrients in soil and improve its general structure and quality, as well as conserving water.

Land can be rehabilitated. Niger underwent periods of severe drought in the 1970s and 1980s that degraded large areas of land. But fortunately some aid money went toward restoring the soil and building water conservation projects. The construction of stone bunds, for example, and the use of traditional planting pits helps save water. More than 300,000 hectares of land were rehabilitated in this way, and over the years both tree cover and crop yields have increased.[41]

༺

I FIND MYSELF THINKING primarily of agriculture as a contributor to climate change, and trying to find ways to reduce its impact. But perhaps just as important is to recognize the impact that climate change is going to have on agriculture.

We saw how the four key crops of wheat, rice, corn, and soybeans provide half the calories grown worldwide. Now look at how the UN Food and Agriculture Organization predict that the yields of these crops will fall under business-as-usual models of climate change. Corn yields would be down by 20 to 40 percent, wheat down by 5 to 50 percent, rice by 20 to 30 percent, and soy yields would be reduced by 30 to 60 percent.[42] These crashes, of course, will be happening just as the world population is going to be peaking, when we need to produce far more food than we are now and when we can't afford to use up more land. And that's not the end of it. As well as getting reduced amounts of crops out of the ground, their quality will also fall if we don't do anything about it. Global heating reduces the nutritional quality of food.[43] People will need to eat more to get the same benefit, which is another unwanted source of demand.

I'm talking about predictions for the future of food production, but it's not as if climate change is not already hurting farmers. Droughts and floods have big impacts on fisheries and agriculture, and these climate change-related effects are increasing in frequency around the world.[44] In the sea, we are past capacity: About 90 percent of industrial fisheries around the world are overexploited or fully exploited.[45]

As well as the hit to farmers' livelihoods, to food prices, and to food's nutritional composition, the food security impact can be significant. For example, in Russia a poor harvest in 2010 forced the country to suspend exports of grain, which led to bread shortages in Egypt and Tunisia, which arguably led to the overthrow of those countries' governments.[46]

So it is essential to prepare our fields, seeds, and fisheries for the difficulties of a climate-changed world. We've seen how sustainable intensification can increase yields. Another, complementary way, is to increase the diversity of crops grown.

An analysis of fifty years of data on yields from ninety-one countries found that a greater diversity of crops grown, at a national level, is correlated with a greater stability of yield. Diversity compensates when you get sudden droughts, smoothing out the crashes in harvests you'd otherwise get.[47] It's an overlooked way to help deliver the yields we need without using more land or fertilizer.

~

THEN THERE ARE SILICON VALLEY–TYPE APPROACHES. In the natural condition, plants don't grow in sterile sponges, sucking up water and nutrients from the soil. They grow in conjunction with dense communities of bacteria and fungi that interact with each other and buffer fluctuations in the environment, such as periods of drought or drops in some critical nutrient, or visitation by a pest species. In conventional agriculture, these fluctuations are smoothed over by irrigation, fertilization, and treatment with pesticides. At the scales we see today, it is unsustainable and wasteful. We don't have the water, and fertilizers and chemicals are overused, contributing to pollution and greenhouse gas emissions.

This is why some agricultural companies are reintroducing microbes to the growing plants. Indigo Agriculture, for example, a biotech company based in Boston, treats seeds with microbes before planting that grow better than regular seeds. In tests in Oklahoma, Kansas, and Texas, microbial-treated seeds gave, on average, a 13 percent higher yield than the normal seeds; under drought conditions the treated seeds produced 19 percent more.[48]

Genetic modification is going to be essential, and Europeans are going to have to get over any misplaced squeamishness about it. There are many crops already available that could provide huge benefits to farmers and consumers, including drought-resistant corn, and golden rice, a genetically modified form of the crop that is enriched with vitamin A. Deficiencies in vitamin A kill more than half a million children a year,[49] yet for twenty years golden rice has been in regulatory purgatory while authorities dither over allowing it to be used; now Bangladesh is about to finally permit its cultivation. A modified banana enriched with vitamin A has also been developed, along with improved apples and potatoes.

Another thing to crack is the C4 Rice Project. The rice plant fixes sunlight using a biochemical process known as the C3 pathway. Other crops, such as corn and sugarcane, use the C4 pathway, which is more efficient. The idea is to engineer rice to use the C4 pathway, enabling farmers to grow more from a smaller area.[50]

At Cold Spring Harbor Laboratory in New York, tomato plants have been gene-edited—genetically tweaked to carry useful genes from other plants. Some plants have a gene that makes them grow their fruit without a kink of material—a knuckle—joining the fruit to the stem. This enables machines to harvest them undamaged, which is not possible with regular tomato plants. Other plants self-prune, so as to take up less room; yet others flower earlier; while some had a higher content of lycopene, a nutritionally important antioxidant. Gene editing is now huge business. The giant multinational organizations Bayer DuPont and Syngenta (now owned by Chinese gene-editing company ChemChina) are all developing gene-edited varieties of many different crops, including potatoes, corn, sugarcane, and soy.

As well as our recipe competition, we'll introduce a climate labeling scheme. Shoppers will be able to see which of the items in the supermarket have a better carbon footprint. Genetically modified crops can be much more environmentally friendly if they can fix their own nitrogen, so require less fertilizer, or have been made drought tolerant, so use up less water. This will hopefully break the stigma of gene-edited foodstuffs in Europe.

⮑

OUR AIM IS TO ACHIEVE A TRANSFORMATION in world agriculture that will benefit billions of people. Hundreds of millions of smallholders will get sustainable livelihoods. Our health will improve. The intense pressure on the planet will be relieved. The unconscionable suffering of billions of animals will, at the least, be reduced. We will be planting for the future—the only future that is possible on a crowded planet.

There's a scene that I sometimes think of from *The Ballad of Halo Jones*, the epic space opera by Alan Moore and Ian Gibson. At one point Halo, a native of a future society, is on an exoplanet far from home where they eat different food than the kind she's used to on Earth. In a canteen she points to some weird gloopy things frying in a pan: "What are these?" "Fried eggs. You want some?" "*Eggs?* What, you mean, from out of some animal's *ovaries*? You mean . . . to *eat*?"

I think of this scene, because one day it will be us. One day people will simply think it's weird to eat animals and the products made by animals. Why not? We are on a trajectory of ever-increasing empathy for our fellow life-forms, of growing awareness of the environmental problems that come from farming animals, and of the sheer wastefulness in terms of resources. A civilized, technologically advanced, healthy, and sustainable society—one that will be able to support ten billion people—will have to move away from the reliance on animals for nutrition. We need to get to a place where we at least think twice before consuming animals or their products. I am under no illusions about the time and effort it will take to get there.

Achieved

A massive reduction in the farming and consumption of animals and an accompanying cut in greenhouse gas emissions. The end to our treatment of the planet as an all-you-can-eat buffet. A curtailment of greed, the end of taking more than our share, leaving the debt to be paid by our children. The transition to a sustainable and humane system of global food production and responsible stewardship of the world's resources. Avoidance of mass famine and ecological collapse.

Money spent

Repurposed hundreds of thousands of square miles of cattle farming for crop agriculture:	$700 billion
Regeneration of lost farmland, soil restoration, improved agriculture:	$100 billion
Development of plant-based substitutes for meat and fish, scaled up for mass production:	$10 billion
Creation of global network of agro-ecological colleges:	$1 billion
Development of low-carbon cryogenic energy storage and refrigeration:	$1 billion
Social capital connecting thirty million groups of farmers:	$100 million
Investment in insect farming:	$10 billion
Sustainable best recipe competition:	$10 million
Climate-smart crops: development and distribution of seeds gene-edited for improved yields and resistance:	$80 billion
Total:	**$902.11 billion**

A supermassive black hole in
Centaurus A, a prominent galaxy in
the constellation of Centaurus.

Discover a New Reality

AIM: To break, or fill the gaps in, the standard model of particle physics. To understand the missing 95 percent of reality. To build a theory of quantum gravity. To map the cosmic neutrino background left one second after the big bang. To move our knowledge of reality.

EACH TIME WE FIND A WAY to look more deeply, we seem to find new things.

I worked for a couple of years in a physics lab at Trinity College Dublin. The group I was with was interested in the structure and behavior of water at a molecular level—you'd be surprised at how weird this most familiar of substances is—and investigated with something called an atomic force microscope (AFM). AFMs are able to probe, like the needle of a vinyl record player with the gentlest of touches and the finest of points, the peaks and troughs made by the actual atoms of a substance, and produce images of its atomic structure. I remember the thrill of seeing a fuzzy image like a chessboard on the screen and being told by the nonplussed physicists that it was the image of the atoms in a sheet of copper. To me, a messy field biologist used to working with things I could see with the naked eye, or at least with a light microscope, it was an incredible thing to see *actual atoms*.

My job was to use a modified AFM to look at the change in body structure of animals that survive dehydration, such as bdelloid rotifers and tardigrades. Both are strange and fascinating microorganisms with great tolerance to extreme conditions, and I wondered what special tricks they might have, at a structural level, that allow them to fold up and sit out long periods without water.

If I was amazed at seeing images of atoms, the grad students in the lab had their minds blown when I got some dry dirt from the gutter outside the lab, added water, and looked at it under a regular optical microscope. The eyepiece showed bits of green algae and chunks of dead plant cells, but also microscopic worms and other "bugs" going about their business, flagellating and undulating in the smear of water; occasionally we'd see a beautiful rotifer twirling past. The physics students had never seen anything like it, and I felt like quite the magician. The beasts I was interested in were the rotifers, multicellular animals about 0.2 millimeters long. That's $2x10^{-4}$ meters. The atoms of the copper sheet I'd seen were about 200 picometers across, which is $2x10^{-10}$ meters. So the difference between the size of the rotifer and the size of the copper atom is about six orders of magnitude. Between me and an atom it's ten orders of magnitude, and between an atom and the tallest building in the world, the Burj Khalifa in Dubai, almost thirteen orders (the Burj is 830 meters high, or about 2,700 feet, but let's round it up and call it 1 kilometer). Remember that for a moment.

Until Anton van Leeuwenhoek got out his microscope, we had no idea at all about the world of life invisible to the naked eye. Using the best light microscopes, we can now resolve images to the very wavelength of light itself, about 10^{-7} meters. If we want smaller than that, we can use electrons instead of visible light, which have a shorter wavelength, so we can see down to 10^{-10} meters. At this scale we are inside atoms, and to see more we have to start smashing them apart in particle colliders. By doing this, we can see inside protons and neutrons, and figure out what is going on at distances of 10^{-20} meters.

Getting to that level of detail has been no easy task. It has taken the greatest and most complex scientific project ever completed, and a monument to human collaboration and ingenuity, the Large Hadron Collider (LHC) near Geneva, Switzerland. We can list some impressive technical details for this. It has ten thousand superconducting magnets in a circular tunnel 17 miles in diameter, cooled by 106 tons of liquid helium, spewing so much data it requires the world's largest computing grid, across thirty-six countries, to process it. But it is one simple point that takes the breath away. The LHC smashes particles together so hard as to rend the fabric of reality. It amazes me that the last bit of that sentence isn't a metaphor: We can break the very building blocks of matter, and measure what happens.

∽

WHEN THE HIGGS BOSON was discovered at the LHC in 2012, at an estimated cost of $13.25 billion,* there was huge celebration and worldwide media acclaim. The existence of the Higgs had been predicted by our best theory of reality, the boringly named (if wildly successful) "standard model." This term was coined in 1975, building on work by many scientists from the early 1960s onward, combining three of the four known fundamental forces

---

* This is slightly unfair, as this is the cost to build and run the LHC, which does other things than look for the Higgs.

of nature: electromagnetism, and the weak and strong forces (but not gravity). The standard model predicted the existence of sub-atomic particles called quarks, and they were duly discovered. With two of the quarks (the so-called up and down quark), you can make protons and neutrons; add electrons, and you have all you need to make any element.

Yet, for all its incredible predictive power, there are several big things the standard model can't explain. It doesn't tell us why there are more particles than are strictly needed to make matter. Nor does it tell us what 95 percent of the entire universe is made of, because it can't explain what dark matter or dark energy is. Nor can it tell us why there is more matter than anti-matter in the universe. It turns out our description of reality is like we've bought a new house and we're telling our friends about it, but we've only actually examined the first room, in the dark, with a penlight. There might be a cellar, a garden, a kitchen, an upstairs—we don't know. It's actually more like buying the house, living for a few years on the doormat, and then becoming dimly aware that there's much more behind the front door. It's a bit em-barrassing to be so ignorant, frankly. Our descendants will look back on us as we do at the ancient Greeks, who thought matter was made of earth, fire, air, and water.

To find out more about reality, we have to look harder. We have to smash particles together with more power. The LHC is doing this, and has been upgraded to enable it to fling protons at each other with even more energy. But there is a *lot* more to see. We know protons are made of quarks, and we know there is another class of particle, called leptons. The electron is a lepton. But is that it? What are quarks and leptons made of?

Physicists know that there is (probably) a fundamental lim-it to how small things can get. The smallest possible length in physical reality is the Planck length: 10^{-35} meters. It's hard to con-ceive how small this is, and how far down you'd have to go to get there. Remember, we're already at 10^{-20} meters, so we need to go down another fifteen orders of magnitude. That's more than the difference between the size of an *atom* and the height of the Burj Khalifa, more than the difference between the size of you and the radius of the solar system. Imagine not knowing *anything* about the processes happening across such a gulf.

Yet this huge range of possibility is currently closed off to us. We may have ways to search indirectly for evidence of the composites of leptons and quarks, but we don't have the capacity in our biggest particle colliders to break them apart to see what's inside, if that is even possible. Not knowing what is there has left physicists with the mother of all headaches, because, while at the same time being proud of the success of the standard model, they really would prefer it if it went wrong.

In 2018, data from the LHC confirmed that the Higgs particle does exactly what the standard model predicts, meaning the model passed its latest and most intense health check. Its health is good—"disgustingly good," said one particle physicist. There is, however, the faintest of cracks in the model in that the neutrino isn't quite what it ought to be. We will investigate this crack in the standard model edifice, because it might give us a clue as to where to look for the missing rooms of the house. But at the moment we have no clue what dark matter is made of, nor do we know why reality "inflated" so insanely fast at the beginning of the universe, nor why the expansion of the universe is accelerating. We ascribe the acceleration to a mysterious dark energy, but we know next to nothing about this force. And we don't know how to describe gravity at a quantum level.

We have an unquenchable desire for knowledge. We *need* to find these things out. The quality of our future depends, too, on improving fundamental understanding. "The answers to the exotic questions of today underpin the everyday technology of tomorrow," says Roberto Trotta, a physicist at Imperial College

London. In the early twentieth century there were breakthroughs in relativity and quantum mechanics, and this led directly to the technology that shapes our lives today. (We'll join the race to build quantum computers in the final chapter.) If we can figure out what dark matter is, and dark energy, and discover a theory of quantum gravity, we might open the door to the long-term survival of our species. In investing in fundamental research, we make no promises of success, but I would argue that there is no other endeavor that impacts so much, not just our understanding of the cosmos, but our philosophy, and our understanding of understanding itself. If we invest our windfall here, we can potentially take a gigantic step forward in human knowledge, reset our entire comprehension of the universe, and accelerate the development of our species.

Science had a long fuse before it got going. The ancient Greeks and Islamic scholars in the Middle Ages did their bit, but science as a framework for investigation didn't really take off until the renaissance in the sixteenth and seventeenth centuries, when there was an explosion of discoveries. To pick a few, we had Copernicus and Galileo putting the Sun at the center of the solar system, Hooke discovering the cell, and Newton describing light and gravity and the laws of physics. The explosion continues today, in all directions. Our job is to feed it.

First in this chapter we will identify the biggest problems in physics, and then we will see what we could do to solve them.

∽

Problem 1: The origin of the universe

With physics, you have to leave your intuition at the door. You just have to accept that what the equations tell us about reality don't marry in any way with our experience of it. That said, there are a few things that just don't seem to make sense, and if you pick at them the problem gets worse. Take the explanation for the origin of our universe.

Astronomers discovered in the early to mid part of the twentieth century that the galaxies were all moving away from each other: The universe was expanding. Originally, some kind of explosion—a big bang—must have started it all off. That doesn't sound too problematic at first, but look at it in the light of Einstein's theory of general relativity. Einstein is talked about in the same way as Mozart, and Shakespeare, and Michelangelo, because like them he created something at the pinnacle of human achievement. In 1915, Einstein showed that gravity was not a force so much as a distortion of space itself; indeed that space and time were in reality merged into a single concept, space-time. I can't *really* understand that, but I marvel at it. Combine the observation that the universe is expanding with what we learn from general relativity, and you conclude that back in time everything was closer together. Everything: the hundred billion stars in our galaxy, and the billions upon billions of galaxies spread out in superclusters stretching for hundreds of millions of light-years. Our supercluster weighs 10^{15} times the mass of our Sun. Further back in time still, all of that was mashed together. At the beginning, in fact, there was a point of infinite density.

As I said, you have to leave your intuition at the door when accepting what the data and the equations tell you, but still—it doesn't compute, does it, the idea that everything, *all the matter in the universe*, was condensed into a singularity of incalculable mass but zero volume.

In a sense intuition is right here. Einstein's general relativity is an account of reality that doesn't explain the weird things that go on at very small scales that are described by quantum physics. This means the description of the singularity at the start of the universe, and of the big bang itself, won't be complete until we know how to combine relativity with quantum physics. One good place to look is where we know there are singularities—at the heart of black holes.

～

Problem 2: The growth of the universe

The birth of our universe is one thing we don't understand. Another is its development as a newborn. If we look out across the universe today, it is both flatter than it ought to be and smoother, meaning it has the same temperature everywhere we look (2.73 kelvin, which is −270°C, or −454°F). To solve this problem, physicists have invoked a concept called inflation: the idea that just after the big bang the universe underwent a period of extraordinary growth, an expansion faster than the speed of light. Now that we're familiar with the sorts of scales physicists work on, we can get a grasp on what we're talking about: Inflation grew the universe by a factor of 10^{25}, in less than 10^{-32} seconds. The speed of light is an unbreakable limit in physics, but it applies only to things *in* the universe. With inflation we're talking about the growth of the universe itself and, incredible as it seems to have such a monumental expansion in such a short time, the equations say it's possible.

This period of growth solves the smoothness and the temperature problem, so physicists are happy with inflation in that sense. But all explanations just throw up more problems—indeed, that's the nature of a good explanation—and the problem with inflation is we don't know what drove it. We need to know what caused the universe itself to grow by twenty-five orders of magnitude in a ridiculously short time. A new kind of force we don't know about? Perhaps the speed of light was itself slower in the early universe, and it's that that is behind the flatness and smoothness.

Even more bizarrely, if inflation does turn out to be the explanation, then it opens the door to the idea of the multiverse.

During inflation, quantum fluctuations in the fabric of the universe cause rifts and bubbles to pop into existence, break away, and start inflating themselves. Multiple universes—an infinity of them, all unconnected—proliferate throughout a greater meta-reality. It's all verified (or at least supported) by equations. The multiverse, by the way, is the concept that inspired Philip Pullman to conceive of alternative worlds in *His Dark Materials*. If only we had a subtle knife to explore them with.* More tangible and more feasible for study is the *rate* of expansion of the universe. It's something else we need an answer to, as we'll see.

~~~

Problem 3: Dark energy

When stars die, many of them blow up. Some kinds of stars do this in a consistent way, and astronomers call them "standard candles" because they allow us to make accurate measurements across the universe. These explosions are called type I supernovae, and in the 1990s two teams were studying them to determine the rate of expansion of the universe. The idea was that either the expansion would slow down and then reverse, as gravity gradually worked as a brake, or that the power of the big bang was such that gravity would not be able to slow it down, and the universe would continue to grow at the same steady rate. The teams found that neither explanation was true: The universe was expanding faster and faster. Something is driving the growth, but we have no idea what it is. It's been called dark energy simply because we need to call it something.

---

* In Philip Pullman's *The Subtle Knife*, there's a scene where Iorek Bynison examines the blade of the titular knife, so sharp as to fade into an atom-thick edge. If you put a carbon nanotube on the tip of the probe on an AFM, you can get a point only a few atoms across. Like the subtle knife, it can help discover new worlds.

It accounts for a *lot* of the stuff of the universe, some 68 percent. But what is it? It could be a kind of energy that exists in the vacuum of interstellar space, caused by quantum particles popping into existence. It could be a new kind of force that we have not yet detected. We don't know, and it's important, if only for our self-respect, that we find out.

~

Problem 4: Dark matter

In the 1970s, Vera Rubin was making observations of the rotation of spiral galaxies, such as Andromeda, also known as Messier 31. At only 2.5 million light-years away, it's the closest one to us. Rubin found that Andromeda and some other galaxies were spinning faster than they were predicted to, based on the estimated number of stars they contained. In the 1930s, a similar observation had been made of the Coma galactic cluster, but no one really knew what it meant and the data were essentially ignored. Rubin's work made it impossible to ignore: Her results showed that there was some invisible matter in the galaxies she was looking at that caused them to spin so fast. The stuff became known as dark matter, and detailed observations over the years since have shown that it makes up 27 percent of the universe.

We know agonizingly little about dark matter. We've seen the epic fallout that has occurred when two galaxies collide into each other (an event that will befall our galaxy and Andromeda in 4.5 billion years), and at the crash site astronomers can see the gravitational distortion that the dark matter makes. We know that massive objects can bend light and, at the site of a collision between two groups of galaxies known as the Bullet Cluster, we see distortions of light caused by the dark matter.[1] The chilling thing is that the visible matter of the galaxies—the billions of stars and planets and dust clouds they contain—cause obvious and measurable explosions as they collide, but the dark matter counterparts of the colliding galaxies just pass through each other. Herein lies the big problem with trying to study it: We can't get hold of the stuff. We know it is under gravitational influence, but it seems to have no response to other forces we know about.

⌇

Problem 5: Antimatter

I had a bet with my father when I was very young. His position was that anything I could imagine from science fiction would come to pass in my lifetime. At the time I must have just seen *Star Wars: A New Hope*, as the first thing I thought of was the holographic projection of Princess Leia. Dad had no hesitation in predicting that this technology would soon be developed. (Sure enough, a few decades later we had Tupac performing on stage from beyond the grave.) "What else?," he prompted. Then I remembered, switching from *Star Wars* to *Star Trek*, the "antimatter engines" that power the warp drive. Antimatter warp engines, I said. That, he conceded, might take a bit longer.

Unknown to both me and my father at the time, antimatter is a real thing. In 1928, the physicist Paul Dirac was playing around with the equations of special relativity and quantum mechanics and found that his combined equation churned out particles with the opposite charge (I've glossed over the details). So, as well as a negatively charged electron, Dirac predicted an antielectron with a positive charge. The same worked for all other particles. Everything had an opposite version of itself: Matter was reflected by antimatter. If a particle of antimatter meets its opposite partner, they annihilate each other, giving off energy in the form of photons (photons, you see, have no mass). Such was the prediction of Dirac's equation, which even he didn't really believe, but then cosmic rays were found to contain particles of antimatter, and now we are able to produce it (with some considerable difficulty) in particle colliders. It has a place in the standard model, but there is a problem: The standard model says antimatter and normal matter were produced in equal amounts at the big bang. So why is there so little antimatter around these days? Certainly there is not enough to power a warp drive.

This is another reason why the standard model is so infuriating for physicists. (Several I spoke to said they wanted to "kill" it.) It has worked well in many ways, but it is clunky, ramshackle, and retrofitted. It is not what physicists most like to say about a theory: It is not elegant. This wouldn't matter so much, but the model also doesn't explain why the fundamental forces have the values they do; for the particles whose existence it so accurately predicts, the model doesn't explain why they have the masses we find. As we've seen, it doesn't explain dark matter or dark energy. The universe appears to be fine-tuned to a spooky degree. If the proportions of dark energy and dark matter were slightly different, galaxies wouldn't form; if the fundamental forces were different strengths, atoms wouldn't bind together, stars wouldn't shine. The antimatter question is the most delicate of these, for if the balance hadn't tipped in matter's favor billions of years ago we wouldn't be here to worry about it.

⁓

IF WE WANT TO MURDER THE STANDARD MODEL, we're going to need a bigger particle accelerator. We'll need to smash particles together at greater speeds than is possible at the LHC. There are plans for new facilities at CERN, but no one seems to know quite what to do at the moment. If new colliders are not built, then physics will have to make do with finessing and tweaking, using smaller experiments to look for oddities that might open up cracks in the standard model. Neutrino experiments, as we'll see below, are one way to go.

For the colliders, CERN is contemplating the Future Circular Collider, which would be a machine of sixty-two miles in circumference, compared to the seventeen miles of the LHC. It would be ten times as powerful, and has been priced at more than $20 billion. CERN also has plans for a linear collider, called the Compact Linear Collider (CLIC). This is similar to something being considered by Japan, the International Linear Collider (ILC) (estimated cost $7.5 billion). There is also a Chinese proposal, the Circular Electron Positron Collider (CEPC). No one is deciding on what to do until someone else makes a move.

One idea is to collaborate, and this is the position we should take with our investment. Machines of this scale and complexity, dwarfing even the LHC, have never been built, and are multi-decade projects. In the real world, there is a fear that we'd invest a huge amount of money and time and even then not find anything new, that we'd only be exploring a "desert" zone. Maybe physics won't give up its next secrets until we get to even higher energy levels. Now that the LHC has started up again at its boosted level, we'll see if we get any hints. That might be what sways investment in the next machines. But, in our ideal world of almost limitless investment, we can press on with the next generation of collider, in a groundbreaking and unifying collaboration between Europe, America, Japan, and China. We *know* there must be new discoveries out there, and eventually we'll have to make bigger colliders to detect them. Our investment will ensure this happens on the fast track (relatively fast: We couldn't start building it right away even if it was fully funded, because we just don't have the technical understanding to build the superconductors it will need, as that will take years to develop). Having found the Higgs, the LHC might not find anything else, even with its upgrades. Even if that is the case, we *will* find out stuff on the way, even if it is about pushing technological boundaries of what is possible.

Some futurists are already thinking ahead, to developing a space-based collider. In space, we might be able to reach energies not feasible on Earth—perhaps high enough to probe matter at the Planck length, which as we saw is the smallest scale allowed by quantum mechanics. With the information generated, we could perhaps detect dark matter. We might uncover as-yet-hidden dimensions. The main problem (of many) is that the technology necessary to construct such an immense machine is not available. We don't even know how to build the successor to the LHC, as we've just seen. The development of everything necessary to even contemplate a large-scale space-based collider—say, one that extended around the orbit of Mars—would cost far more than world GDP. This is something for the far future. Better we stick with land-based colliders for now. Though, as we'll see, we might want to put one on the Moon.

∽

ONE PROPOSED SOLUTION for the dark matter mystery is the axion, a hypothetical particle that might comprise the missing matter. If this elusive beast exists, we might be able to force it to turn into a photon, which we could then detect. The existence of the Higgs was deduced in a similar way, by tracking the debris thrown off when a boson was smashed to pieces. There's one detector designed and waiting to be built in Germany, MADMAX, that could perform this experiment. (It deserves funding for the name alone.) We can pay for other axion experiments that are in the planning stage—Orpheus in the United States, and CUL-TASK in South Korea—to look for dark matter.

Another candidate for dark matter are the WIMPs: weakly interacting massive particles. The hope would be that these or axions would be detected in the next generation of colliders.

Scientists are also keen to test supersymmetry, the idea that each particle in the standard model has a "partner particle" of sorts. And, of course, there is the matter–antimatter problem. To wit: If matter and antimatter were created in equal amounts, they would have annihilated each other, so how did the matter that makes us manage to survive?

The investigation into neutrinos is something that we can pursue without the need for a new collider. Neutrinos are odd, ghostly particles that have a place in the standard model list but about which we are unsure, because they hardly ever interact with other forms of matter, despite being some of the most abundant particles in the universe. They seem almost immune to interaction, and theoretically they could pass through a wall of lead trillions of miles thick without touching an atom. They are so gossamer as to barely be there—the lightest of all the known subatomic particles that have mass, at least six million times lighter than an electron.[2] Many billions of them are passing through you and me right now, right through the entire Earth and out the other side. They seem to wink in and out of existence and somehow change their nature. We need to know more about them, because their behavior might enlighten us about some of the mysteries of the standard model.

Some 75 miles north of Tokyo on the Pacific coast, in the otherwise sleepy village of Tokai in Ibaraki Prefecture (it's not far from where I worked as a postdoc), sits one of the world's most advanced particle accelerators. J-PARC, the Japan Proton Accelerator Research Complex, has a pointy end, unlike the circular accelerator at CERN. It's more like a gun, in that it generates neutrinos, and fires them through the bedrock of Japan to a detector deep under a mountain named Kamioka, around two hundred miles away. The origin and target of this neutrino beam—Tokai to Kamioka—give the experiment its name: T2K.

Decades ago, the Mozumi Mine under Kamioka mountain used to supply zinc, silver, and lead to Japanese industry. Now it is home to the Super-Kamiokande detector, a vast stainless steel tank filled with 55,000 tons of ultra-pure water. Surrounding the tank are more than 10,000 photodetectors. The whole setup is exquisitely designed to detect the interaction of a neutrino with the atoms of water in the tank. It has to be housed deep underground so that the rock shields it from cosmic rays bombarding the surface of the Earth. When a neutrino interacts with matter, a tiny flash of light is generated, and this is what the photodetectors are set up to catch.

Neutrinos are elementary particles that come in three types. Physicists refer to them as flavors in a roughly analogous way to how biologists might describe birds as belonging to different species. Known as electron neutrinos, muon neutrinos, and tau neutrinos, they are also incredibly slippery. Neutrinos are electrically neutral and have barely any mass, so they hardly ever interact with other forms of matter; they just pass through almost everything. But we do know they can change flavor. It is the *way* they do this that might shed light on the question of why there is more matter than antimatter in the universe. It might tell us about the entire "dark sector" of reality, the shadow land of opposites about which we have only the faintest idea.

But, to really find out, an international collaboration of physicists is already planning the successor to Super-Kamiokande, the Hyper-Kamiokande. Fifty-five thousand tons of water is not enough, it turns out. Hyper-K will contain 1.1 million tons. As with the colliders, there are alternative proposals. An international collaboration has plans for the Deep Underground Neutrino Experiment (DUNE) to be built in the US. Neither Hyper-K nor DUNE are yet being constructed, so we will immediately provide funding for a joint venture.

AND THEN THERE IS THE PHYSICS DISCOVERY of the decade: The detection, in 2016, of gravitational waves. These are waves in the very fabric of space-time caused by the movement or collision of massive objects, and were detected by the Laser Interferometer Gravitational-wave Observatory (LIGO) based in Italy.

Gravitational waves are a huge deal, not just because they were predicted by Einstein but because they mark a sea change in what we can do with astronomy: Instead of relying mainly on light, as we have for four hundred years, we can now observe the universe using gravity. It means we can study things that are invisible to electromagnetic radiation, such as black holes and dark matter and the big bang. We might well discover exotic kinds of stars that are currently only theoretical: stars made of quarks, perhaps, or preons (a hypothetical particle that is a level below quarks).

By observing the gravitational ripples, cosmologists will be able to explore fundamental physics in ways never before possible. The only means we currently have to investigate dark matter is through its effect on gravity, as it simply has no interaction with any other force or particle we know of. We can also hope to find out more about the nature of dark energy. Remember, this is the stuff that is causing the universe's expansion to accelerate.

But, to investigate these mysteries, we will need to build more sensitive gravitational wave detectors, and that means doing so in space. ESA has plans for such an observatory. It is called LISA—the Laser Interferometer Space Antenna mission—but is only slated for launch in 2034. We can accelerate the era of gravitational wave astronomy.

GALILEO DIDN'T INVENT THE TELESCOPE, but his observations changed the world. Since then, with every improvement in optics and telescope design we've learned more. In Chile, for example, the European Extremely Large Telescope (ELT) is currently being built at altitude. Astronomers will use it to measure the speed of expansion of the universe accurately, and get an idea of how this might have changed with time.

Telescopes are not cheap or easy to make. The successor to Hubble is the James Webb Space Telescope, but the project began in 1996, finally cleared for launch in December 2021, and its cost has risen from a projected $0.5 billion to an estimated $9.6 billion. The James Webb is an exoplanet hunter, and will allow us to look at the atmospheres of alien worlds (as we saw in chapter 6).

Mountaintops are good for such telescopes because the atmosphere is thinner, meaning the stars twinkle less (twinkling is caused by turbulence in the atmosphere). But it is still not ideal. Plus if we want to do radio astronomy, and measure the radio waves emitted by distant objects, Earth is a bad place because it is swamped by radio pollution from satellite transmissions, mobile phone signals, digital TV, and the like. Low-frequency radio waves that might carry faint signals from the big bang get screened out by the Earth's ionosphere, so we need to investigate them somewhere free of radio transmission interference, and the best place in the solar system is the far side of the Moon. It never faces Earth, so is permanently shielded from radio noise.

Some of the strongest evidence that the big bang occurred in the way cosmological models predict is the detection of the oldest light in the universe. For a few hundred thousand years after the start of the universe, the heat was so intense that everything existed as a dense hot plasma fog of protons and electrons. But, when the universe was about 380,000 years old, it cooled enough to allow the formation of atoms. This was like when a morning fog clears and the view out the window becomes clear: The universe became transparent. Some space-based instruments have mapped the pattern of this ancient light to form what's known as the cosmic microwave background, an image of what the universe looked like before stars and planets had formed. Studying this map reveals kinks and lumps that show variations of gravity that are the seeds of future galaxies, and we can make estimates of matter, dark matter, and dark energy.

A Moon-based radio observatory would allow us to test theories of inflation and understand the early years of the universe. We'd use robots to deploy millions of radio antennas over the Moon's far side. We can also build infrared telescopes in craters on the South Pole. These are some of the coldest known parts of the solar system, having never received the light of the Sun.

But we could also do something even better. We looked earlier at neutrinos, the ghostly particles streaming through the universe. Like the cosmic microwave background, neutrinos also separated from the rest of the hot matter of the universe early on and left an imprint, but they did it when the universe was only *one second* old. We now have evidence that this cosmic neutrino background exists; if we could make an image of it, we would get a new handle on measuring dark matter and inflation.[3]

∽

THESE ARE SOME OF THE BIG PHYSICS EXPERIMENTS being planned and being built—and some that are being dreamed of. It is a long game. It takes decades to plan and construct machines on this scale. Like LISA, we will need to build some observatories and experiments in low-Earth orbit, or on the Moon, to escape the various sources of interference and contamination that we find on Earth. Indeed, it's only by building a science base on the far side of the Moon that I've been able to put a dent in the $1 trillion. That's a sobering and frustrating discovery. There is so much we could be doing for relatively little cost, even without going to the Moon. Yet, for lack of funding, projects get mothballed, or delayed for years.

The problems are so big, so basic, and so intimate that we need to look deeper and harder at the foundations of reality than ever before. I don't think it contravenes the rules of this game if we spend a chunk of the money on fostering global research and education initiatives. Physics, like much of science (and society), has a diversity problem, which we can address by creating opportunities for women and building research institutes in Africa, Asia, and South America.

Look at what we can do compared to the ancient Greeks. We have fMRI machines that can look into the body and see the workings of a living brain; we can land robots on asteroids in deep space; we not only know that matter is made of atoms, we can entangle them using our knowledge of quantum physics. The equivalent advances we might make if our understanding of reality moves on are unimaginable, because we simply don't know what we might find. Arthur C. Clarke's aphorism comes to mind: "Any sufficiently advanced technology is indistinguishable from magic." A world of magic and extraordinary wonder awaits, and we have the chance to get there faster.

Achieved

Alas, we cannot predict what we will discover, but all the scientists consulted expect these projects to reveal physics beyond the current standard model. We can hope and expect to discover the axion, the dark matter particle, and to understand dark energy. We might hope to unify quantum physics and relativity. But there are no guarantees.

Money spent

Hyper-Kamiokande: . . . . $1 billion

Future Circular Collider: . . . . $20 billion

International Linear Collider: . . . . $10 billion

Gravitational waves detector in space: . . . . $10 billion

Science base on far side of the Moon: . . . . $120 billion

Miscellaneous other Big Physics experiments: . . . . $100 billion

Network of research institutes around the world: . . . . $100 billion

Total: . . . . **$361 billion**

The android Junko Chihira developed by Toshiba speaks Japanese, Chinese, and English. It is based at Aqua City shopping mall in Odaiba, Japan, 2016.

10

Second Genesis

AIM: *To develop a machine with human-level intelligence—a machine we would agree is as conscious as you and me. To create a new life-form by assembling an organism with a genome of entirely synthetic genetic material, and endowing it with functions not seen in the natural world.*

HEPHAESTUS IS NOT THE MOST FAMOUS of the ancient Greek gods, but he is surely the most modern. The god of metallurgy and the forge, innovation and technology, he was cast from Mount Olympus for the "crime" of bearing a deformity (a lame foot). Though shunned by the other gods, he was revered for his unparalleled skill in manufacturing. You'll have heard of the winged helmet and sandals of Hermes, the bow and arrow of Eros, and the armor of Achilles. All were crafted by Hephaestus. He also made a replacement shoulder blade for Pelops, the king of Pisa.

But his greatest talent was in automation. We meet Hephaestus in the *Iliad*, when Thetis visits him in his forge to commission a suit of armor for her son, Achilles. Thetis sees that Hephaestus has constructed an extraordinary fleet of autonomous wheeled tripods. Reading this now, it could almost be a description of the Waymo self-driving cars trundling around Silicon Valley. He basically preempts Google's and Apple's autonomous vehicles by some three thousand years. Thetis marvels that Hephaestus even has a staff of automata, "fashioned of gold in the image of maidens"; builds Talos, a giant bronze war-droid that guards Crete from invaders; and is commissioned to create Pandora, the beautiful android sent by Zeus to punish humans for the theft of fire. Homer and the people who dreamed up Hephaestus saw into a future we're still only just reaching, with autonomous vehicles and an artificial intelligence capable of passing the Turing test.

You can see where I'm going with this. We're going grand. Let's do in reality what Hephaestus did in mythology. Let's create a new life-form. Dreams of automation have been with us for a long time, perhaps for as long as we've been making tools. Dreams of *creating* life probably for almost as long. Both are now the goals of multi-billion-dollar industries, and the basis of a global race for dominance. They rest on two of the greatest and most ancient questions that people have asked. What is consciousness? And what is life? It is these that we will tackle here. They are mighty questions, but slippery, and we'll come at them piece by piece.

We've invoked Hephaestus, but we won't copy him; we are bound by the rules of Project Trillion: Military spending is banned, and we want to see tangible results. Hephaestus, the outcast, followed the orders of the more powerful gods. Pandora was a device designed by Zeus to deliver misery, pestilence, and old age. No matter that Pandora's box also contained hope: It was overwhelmingly bad news for humanity. We do not want to unleash the catastrophe many fear could unfold if artificial intelligence goes wrong. Our rules are to ensure that we spend money to reduce human suffering, to increase scientific knowledge, and to benefit the natural environment.

In the latter part of this chapter we'll look at creating a biological life-form from scratch, which will mean tackling the question of what life is. First, though, we'll join the race to build a digital, computer-based entity capable of human-level flexible thinking.

～

OUR GOAL IS WHAT IS CALLED artificial general intelligence (AGI). The *general* is the thing. There are accomplished AI systems already in operation, but their skills are nontransferable.

One of the world's leading AI firms is DeepMind, which is owned by Google. It created a computer program called AlphaZero, which became the greatest chess player of all time when it was given the rules of the game and played itself over and over again, hundreds of millions of times. AlphaZero is phenomenal, a breakthrough AI, but it can't tell you if it looks like it's going to rain. A weather program might register the presence of clouds and predict rain, but that's not because it knows rain comes from clouds, but because it knows clouds are statistically associated with rain. It doesn't have common sense or, as computer scientists say, causal inference. Watson, the IBM AI that beat the best humans on the quiz show Jeopardy! can't play a six-year-old at chess, let alone AlphaZero.

Other AIs can play video games, or identify pictures of cats, or recognize your voice and answer questions, or drive cars, or guide missiles to enemy targets, or choose films they think you might want to watch, but none have flexibility to transfer their abilities to new tasks. This is what people mean by AGI, which is also known as human-level AI. It doesn't mean an AI that will be *as* good as humans at certain tasks; a machine with AGI will far outstrip us, *and* it will be able to transfer its skill to novel challenges, *and* it will have common sense.

If we had a system with flexible intelligence, we would be able to deploy it to take care of a wide range of different tasks, in medicine and social care, across industry and manufacturing, in science and design, in defense, in transportation, in food production, in space exploration, in green energy generation and consumption. In many areas, in fact, that we've spent money on in earlier chapters of this book. If we had AGI, many people think it would unlock unprecedented economic growth and wealth creation, even as it disrupts our society in ways that would be unthinkable if we hadn't just experienced a disruption wrought by a devastating global pandemic. This is why there is a headlong dash to develop it. The reason *we* should invest is that our cash can make sure the fruits of the innovation are shared, and benefit everyone. We're also interested in the biggest questions in science, and they don't get bigger than this.

We should first address the Skynet (Terminator) scenario: the idea that a superintelligence will end up wiping out human civilization. In 2016, Stanford University launched a century-long series of reports into AI, the One Hundred Year Study on Artificial Intelligence (AI100). Their first report concluded that AI does not pose a threat to humanity, at least in the near term. The Stanford panel did say, however, that if the economic benefits of AI are not shared across society, and if the changes that come about are viewed with mistrust, then essential work to ensure the safety of AI technology will be hampered. The report finds that, at least up to 2030, AI, while certainly disrupting some industries, will overwhelmingly provide economic and social benefits. It is our job, while keeping an eye on the post-2030 world, to ensure that the benefits *are* shared.

Some AI research teams have pledged to do just that. OpenAI is a San Francisco-based firm set up to develop human-level artificial intelligence and to try to make sure the benefits are spread out fairly. In 2019, they attracted a $1 billion investment from

Microsoft, most of which will be used to buy time on Microsoft data farms. Data processing—the learning time used to train AIs—is immensely absorbent of processing time. Greg Brockman, the CEO of OpenAI, said the $1 billion would be burned away in under five years. We could invest a large sum, say $50 billion, to allow AI developers around the world more data processing time, in return for a commitment to sharing results and being transparent about what we're developing.

~~

You might be wondering what the fuss is about, if you think of AI mainly as something that can play chess very well. So let's look at some examples of what AI can already achieve, as a hint of what we might expect. First, in medical diagnosis.

One in every three diagnosed cancers is a skin melanoma. Globally, they occur with the greatest frequency in Australasia, North America, Eastern Europe, Western Europe, and Central Europe, and these are also the regions with the greatest mortality and losses to well-being known as disability adjusted life years (DALYs). One DALY is equivalent to one lost year of healthy life. A survey of illness carried in 2015, the Global Burden of Disease Study, found that there were over 350,000 cases of melanoma, more than 1.5 million DALYs, and almost 60,000 deaths.[1] According to the World Health Organization, one in five North Americans will develop skin cancer in their lifetimes.

Now consider diagnosis and treatment. If you suspect you have a melanoma, you typically have to book an appointment at your nearest clinic and have an examination; you may have to then get a referral to a specialist. That's fine if you discover the suspicious bit of skin soon enough, and get the correct diagnosis. But for many thousands that's not the case. So Andre Esteva and colleagues at Stanford University turned to automated classification, having a computer learn from nearly 130,000 images of skin lesions, each one labeled with the name of the underlying disease. Once trained, the algorithm and two dermatologists were tested with novel images of skin lesions, some benign, some malignant. The computer correctly identified the conditions with a 72 percent accuracy, compared to 66 percent for the dermatologists.[2] In more rigorous testing, the algorithm matched the diagnostic accuracy of twenty-one dermatologists. For Esteva, this is a diagnostic tool that should eventually break out of the lab setting and get into the world. All you would need is an app on your smartphone, and you could get universal low-cost diagnosis.

There are many other examples in health care. AI can scan images of the retina and predict the risk of eye and cardiovascular disease; other algorithms can detect breast cancer from mammograms. Some systems are taking the role of the general practitioner. Kang Zhang's team at the University of California in San Diego trained an algorithm on detailed medical records from 1.3 million children who attended a medical center in Guangzhou, China. The records comprise medical charts, lab tests, and doctors' notes. Basically the algorithm soaked up all that knowledge, as if it had itself seen those 1.3 million children. When it was tested on new cases that it hadn't encountered, it was able to diagnose a range of diseases—glandular fever, flu, chickenpox, hand-foot-and-mouth disease—with between 90 and 97 percent accuracy. In tests, the AI outperformed junior pediatricians, and was beaten only by experienced doctors.[3] With more data, the algorithm will only improve; Zhang is now training it on adult illnesses.

No one is saying (yet) that we don't need human doctors, but given their workloads, and the patchiness of global access to them, it seems that algorithms could free up a lot of their time. When health care systems become overloaded—as happened with the coronavirus outbreak—a reliable AI that could answer people's calls, diagnose symptoms, and dispense advice would be not just invaluable and convenient, it would be lifesaving. Indeed, a reliable AI that had data on infections and deaths might have provided advice to prevent the virus from going pandemic in the first place by enabling better tracking and tracing of disease cases, and by modeling different public health interventions and letting them play out virtually. But collaboration between epidemiologists and AI researchers only really got going after the pandemic took off. We will need to continue to nurture this relationship so that we are better prepared for the next pandemic.

HUMAN DOCTORS WILL BE AROUND for the foreseeable future, because their role can be augmented, supplemented, and boosted by AI, but not replaced. And, because dealing with a human may be preferable for those life-critical health discussions we will all end up having. But, for many of our jobs, it might not be worth keeping humans in the loop. Taxi driving. Reception desks, help desks. Long-distance haulage. Assembly line construction. Food preparation. Farming. Human resources. Unilever has said it saved 100,000 hours of human labor in 2019 by screening job applicants by AI rather than face to face.

Let's not underestimate the concerns. The EU is looking at a system to deploy at borders to detect evidence of deception on the faces of people trying to enter the Union,[4] but we know that face recognition systems perform less well when trying to classify non-white men, for example. AI can do things with face recognition that we can't. A team at Stanford University demonstrated one problem by training an AI to determine someone's sexual orientation from their facial features alone. The system correctly identified gay and heterosexual men from single photographs in 81 percent of cases, and in 71 percent of cases with women.

Humans trying to classify the photos managed 61 percent accuracy for men and 54 percent for women. The researchers, who obtained the photos from public profiles on a dating site, say that their work highlights the threat that computer vision algorithms could pose to the safety and privacy of gay people.[5]

That's an illustration of the sort of tension between how AI may help or hamper us as a society. In other areas, AI is more unambiguously helpful. In 2016, DeepMind designed an AI to analyze the efficiency of the cooling system used in Google datacenters. These gigantic facilities handle all the Google searches made in the world, all the YouTube videos being watched, all the Gmail data that has been stored and sent. They are fed by dedicated power stations, which pour vast amounts of energy into the server farms and require a complex cooling infrastructure to keep the temperature down. The AI made recommendations to improve the performance of the system, and these led to a 40 percent drop in the energy used to cool the facility. Google now allows the AI to run the system itself, rather than make recommendations that are then implemented by humans.[6] The company wants to explore how the AI can be used in other industrial settings, delivering energy savings and helping reduce carbon emissions.

In a similar way, AI could help run the electricity grid more efficiently. Nothing will save us other than stopping burning fossil fuels, but advanced AI could, the cheerleaders say, be a key tool in the battle to bring down carbon emissions and tackle the climate crisis.

If we are to invest in AI, we should be open and transparent about what we're doing and work to avoid known pitfalls, dangers, and biases, such as those in face recognition. Data security is another serious issue. DNA-sequencing firms currently sell data to biotechnology and pharmaceutical companies, for example. If you've submitted your DNA for ancestry determination, or for sequencing, it may be that your genetic information has been passed on to secondary companies for analysis. DeepMind was reprimanded when it was revealed that it had used sensitive medical information about 1.6 million people registered with the UK National Health Service without those people's informed consent.[7] One way to make things more secure could be to use blockchain technology—the same thing that underpins Bitcoin cryptocurrency—to help protect patient data. The more this encourages people to give access to their medical data, the better algorithms will become at interpretation.

∽

THIS IS NOT TO SAY that we should create an all-powerful AGI, then prostrate ourselves before it to sort out our lives and clean up our mess. That's probably not a great idea. For one, we might not want a single entity to have this much power over society. For another, despite what it's doing in Google's server farms, AI is far from ready to take on broad advisory roles. Nor should we trust what companies and governments might do with AI power. This is why our investment must be very clearly pledged to be open, fair, shareable, and transparent, with developments and ethical implications discussed and debated.

Some of our investment, then, should go into research in AGI, and part of that will be money spent on processing time for machine learning. There's an issue here that we should be aware of. Moore's law—the doubling of transistor number and therefore processing power in computer circuits every two years—will not go on forever, and indeed may be coming to an end. But the thirst for computational power will only increase.

One solution is quantum computing, which uses algorithms that exploit the hideously unintuitive properties of quantum physics, such as entanglement. In a nutshell, a regular computer uses bits—0s and 1s—to encode information. So it's *either* a 1 *or* a 0. Quantum computers use quantum bits ("qubits"), which can be many different things at once. So in theory quantum machines can store and process information far faster than classical machines.

In practice this is only just starting to happen, because it's been technically difficult to build quantum computers. But the field was buoyed in 2019 when Google announced it had achieved "quantum supremacy," meaning its quantum computer had solved, in 200 seconds, a mathematical problem that would have taken even our best regular computers thousands of years.[8] The feat was compared with the Wright brothers' first flight, as similar world-changing repercussions are expected from quantum computing as have been seen with air travel.

What, then, if the almost supernatural processing power of quantum computing was paired with the form of AI known as machine learning, which is at the basis of most of the examples of AI we've discussed so far. Machine learning is computationally expensive; it's why OpenAI will spend $1 billion mostly on data processing. If we could train algorithms using quantum computers, we could do it faster, more cheaply, and more efficiently than we do at the moment.

As yet, making quantum neural networks for deep learning is only an emerging field.[9] There are formidable technical barriers. But we can get in at the start. Europe, the US, and China are each putting billions into developing quantum computing. Europe has the Quantum Technologies Flagship project, and China, which wants to pull clear of the US in the race for quantum supremacy, has just opened the National Laboratory for Quantum Information Sciences. As with space travel, our aim should be to create a well-funded umbrella organization that fosters collaboration between these competing partners. Let's call it the Tangled Bank, a nod to a key weird aspect of quantum physics, and to the poetic final paragraph in Darwin's *Origin of Species*, where he considers a tangled bank on a riverside and its mass of evolutionary potential.

∽

WE SHOULD INVEST IN WORK toward both artificial general intelligence and quantum supremacy. Even if we don't get to AGI, and we shouldn't underestimate how tricky a problem it is, we'll get lots of benefits on the way. A deep question here is whether a computer with human-level intelligence needs to be conscious. I think working on AGI is helpful for this question because it forces us to get to the heart of what we mean by consciousness. Part of the reason there's so much confusion around the subject is that we all know implicitly what is meant by consciousness, but go all vague when asked to be specific.

The biggest deal for the last few decades has been figuring out the "hard problem" of consciousness, but perhaps the reason no one has resolved it is that it hasn't actually been stated in a way that allows scientists to tackle it. The hard problem as stated by philosopher David Chalmers is "why should physical processing give rise to a rich inner life at all?"[10] In other words, why do we experience things? Why should the color of a red rose or the smell of coffee—qualia, as philosophers term these things— trigger subjective *feelings*?

Keep that in mind for a minute while we consider what qualities an AGI will need. A computer with human-level intelligence must be able to plan and strategize for the future. It will need the ability to use short-term memory. It will need to be able to model what its actions might do, and to understand the actions of other agents. An AI will need theory of mind: the ability to understand the motivation of others and to interact appropriately. If we see a dog off its leash running toward a child holding a sandwich, for example, it will instantly run through our minds that the dog might snatch the food from the child. The child will then be upset or even hurt, and we see that the child's parent, staring at a phone, is unaware of the drama. We can effortlessly process this scene, and probably even my description has you imagining the possible outcomes. For a computer, this is a daunting and complex and mostly unintelligible situation. The way that most scientists think we'll replicate this kind of thinking with computers is through machine learning.

Machine learning is loosely modeled on how we think the brain works. It commonly uses software that describes a complex network of connections, in a similar way to how neurons connect with each other in the human brain. Rather than program all the possible outcomes into the software—which is what software engineers used to try to do, with inevitable shortcomings—in machine learning with a neural network, the computer learns on its own. There has been spectacular success with a turbo form of machine learning called deep learning; it's behind the ability of DeepMind's AlphaGo and AlphaZero, and it's the basis of a system developed by OpenAI called Generative Pre-trained Transformer, or GPT.

A publicly available version called GPT-2 can generate original text, perhaps a sports report, a movie review, or maybe even poetry, when given a prompt. It is a kind of neural network that relies on what's called unsupervised learning. That is, it has been exposed to lots of data (in this case, some eight million text documents scraped off the internet), but like AlphaZero had to learn chess by itself, GPT-2 had to figure out what it all means by itself. It's done a pretty good job. I gave it the first line of this chapter: "Hephaestus is not the most famous of the ancient Greek gods, but he is surely the most modern" and the algorithm continued: "He is first called the Prince of Light and is identified as the guide of the gods and the ruler of the people of Olympus. What is more,

Hephaestus is also associated with the fabled form of the Babylo-nian cult of Anubis." I prefer my opening, thankfully—and GPT-2 is wrong about Hephaestus being the ruler of Olympus, everyone knows that is Zeus—but this is only the basic, publicly available version of the language-generator. More advanced versions can construct impressive arguments, if given enough prompts. I might have asked GPT-2 or its advanced sibling, GPT-3, to make a comparison between what Hephaestus did in mythology and what scientists are trying to do with artificial intelligence, and it might have come out with something more like what I ended up writing.

⌇⌇

THE DAY WHEN DEEP-LEARNING ALGORITHMS are able to write books is a long way off. But there's a lot they will be useful for. They can already search libraries of information far more efficient-ly than people. The amount of information we are generating is almost unimaginably vast. Here's just one example. We mentioned the new radio telescope, the Square Kilometre Array, in chapter 6. It will produce 1 petabyte (that's 1 million gigabytes) of data per *day*. No human could possibly assess it. An AI can devour it.

It reminds me of the Library of Babel, as conceived by Jorge Luis Borges, which contains all possible books. Most are non-sense, but somewhere in the vast library are books that perfectly describe the future, that contain true complete biographies of ev-eryone, even those yet to be born, as well as millions more false biographies. In every language. The librarians search mostly in vain their entire lives for books that contain even one meaningful sentence, such is the size of the pool they sift through.

But what if there was a computer that could search the library at lightning speed? Deep learning can be that speedy librarian. Certainly in chemistry it is already performing that function.

The "possibility space" for different potential drugs is huge, with around 10^{60} different potential drugs that could be made. That's more than the number of atoms in the solar system. But deep learning has been used to search this space, this "chemistry of Babel," and discover a powerful new kind of antibiotic, a class of drug that we desperately need more of.[11] The researchers, with a healthy sense of irony, named the AI-discovered drug "halicin," after HAL from *2001: A Space Odyssey*.

AI with a working memory, able to apply something learned in one context to use in another, has been demonstrated at Deep-Mind. Complex reasoning is one of the things that humans can do as a matter of course. It means we can respond correctly when we are presented with statements such as "Replicants are afraid of Blade Runners. Rachael is a replicant. What is Rachael afraid of?" Or we can look at the London Underground map and tell someone how to get from Old Street to Putney. It's something that computers have had trouble with, but that DeepMind is starting to tackle with a neural network-style computer that has access to a short-term memory.[12] It's a small step toward human-like thinking, and it's this sort of success that encourages DeepMind that the neural network approach is the route to get there. AI with theory of mind isn't far away.

We've looked at a lot of different AI abilities. What if we are able to make an AI that can put them all together? We'll have something that can use knowledge from millions of hospital records to diagnose illness better than human doctors, and certainly without the waiting time; that can design powerful new pharmaceuticals and antibiotics; that can strategize, determining the most efficient way to run a nation's power system; that can drive us around more safely than human drivers and make decisions on what to do when a dog runs out into the road; that can recognize and interpret our emotions, empathize with us, answer our email in the style we usually write in, make phone calls for us, entertain us, play with us. If an AI can do all these things to a high level—planning, using short-term memory, predicting what its actions might do, understanding the motivations of others, being creative—and it seems reasonable to think that we will eventually develop such a machine—then who cares if it "experiences" the smell of coffee or the color of a rose? Its common sense, derived from its deep-learning algorithms, might be effectively as unquantifiable as ours.

In Philip K. Dick's *Do Androids Dream of Electric Sheep?*, Deckard, a bounty hunter known as a Blade Runner, interviews Rachael, a woman who may or may not be an artificial human. During the interview, he measures her physiological reactions.

"You are given a calf-skin wallet on your birthday."

"I wouldn't accept it," Rachael said. "Also I'd report the person who gave it to me to the police."

Deckard tries again, telling her about a time in the barbaric past when people cooked lobsters by dropping them live into boiling water.

"Oh god," Rachael said. "That's awful! Did they really do that? It's depraved! You mean a *live* lobster?"

Deckard notes that, verbally, she has responded in a normal human way, but the instrument he is using to measure her physiological response doesn't register any change: "Formally a correct response. But simulated." Rachael, Deckard concludes, doesn't exhibit *real* empathy. This is the factor, in the book, that separates humans from androids. But the picture is muddled. Deckard himself doesn't show much empathy, certainly not to the androids he kills. And it leaves aside the question of what difference there is, if any, between simulated responses and "real" responses. Our sympathies are with the replicants, but does it matter if it's almost impossible to tell the difference? I'm not so sure.

I'm leaning toward agreement with the idea that we could simulate a conscious being, with "common sense" and human-level intelligence. But unfortunately I don't think it's going to happen any time soon. I say unfortunately, because I'd love to meet an intelligent alien and this is the only likely way it will happen. The CEO of the Allen Institute for AI, Oren Etzioni, says we are still years away from even recreating the intelligence of a six-year-old, let alone full general intelligence. How many years? "Take your estimate, double it, triple it, quadruple it. That's when," he has said.[13] Our job is to help us along the road and to be open about it. Even if we don't reach our destination, the journey will be immensely informative.

❦

As well as his autonomous machines, Hephaestus made a fluid that conveys the quality of life, and it's this small matter that we turn to now. Just as trying to make an intelligence machine shakes up our understanding of what consciousness is, so trying to make organic life will change how we see life itself.

It's probably easier to create an organic life-form than it is to make a truly intelligent machine, because it is harder to explain what the human brain does, and how it does it, let alone mimic it, than it is to create a living system. That's because with living systems we understand a lot more about their components and we know quite well how they are put together. The brain, by contrast, is the most complex object we know of in the universe.

In principle we can approach the creation of a living system like we would the construction of a computer or a highly complex system that is put together with inert matter, with silicon and wires and superconductors; we can use the same approach, but with living matter, and engineer a living organism. This is the goal of synthetic biology: to produce a set of engineering and design rules, and to create microscopic "off-the-shelf" modules for different components of the cell, which will allow us to build a living system. Let's manage expectations right away, however. We are a long way from complete or even detailed understanding of how all the components of a cell work together. We *are* making progress, but we won't be creating complex animals—rather, single-celled organisms on the scale of bacterial cells and, perhaps, yeast cells (which are vastly more complex).

We are pretty sure it can be done. A good proof of concept came in 2010, when geneticist Craig Venter's team assembled the genome of a simple bacterium from synthesized chunks of DNA, and inserted the DNA into the emptied structure of another bacterium. The team, which included microbiologists Hamilton Smith and Clyde Hutchison III, then effectively rebooted the system—and the cell came to life.[14] The genome was that of the bacterium *Mycoplasma mycoides*; the empty host cell it occupied, and then took over, was a related species, *Mycoplasma capricolum*. The rebooted cell grew and multiplied as a normal *M. mycoides* cell.

Venter's group then spent the next half dozen years gradually stripping down the synthetic genome until they were left with only the essential genes necessary for life. At least, life as that cell knows it. That minimal genome was made of 473 genes, almost a third of which (149) were of unknown function. The complexity shocked the team, who had expected a minimal genome to be composed of fewer genes, and to at least know the function of more of them. It is a taste of the difficulty of the whole field of synthetic biology: Everything is much more complicated than you imagine.

$$\sim\!\!\sim$$

A BACTERIUM IS ONE THING, but what we'd prefer to create is a complex cell. Bacteria are prokaryotes, simple cells without complex internal machinery. We—and all other plants, animals, and fungi—are eukaryotes, made of bigger, more complex cells able to perform many different functions. Eukaryotes are more stable, more reliable, more robust, and more programmable. Making them from scratch will be that much more difficult.

Yeast is one of the more simple of the eukaryotes, being a single-celled organism. It forms colonies, but the cells are all the same, not differentiated like they are in frogs and daisies. *M. mycoides* has 473 genes; yeast has 6,275, formed by more than twelve million base pairs of DNA spread across sixteen chromosomes. An international consortium set up to make artificial yeast, the Yeast 2.0 project, doesn't aim just to synthesize and reconstruct these chromosomes, but aims to *rationalize* the entire design of the organism. The new yeast will be cleaned up and optimized, the messy stamp of natural selection removed, repetitive elements taken out. The team has given it an entirely new chromosome, with new functions.

Yeast offers far more versatility, sophistication, flexibility, and reliability than bacteria; if they were modes of transportation, bacteria would be a wooden cart, while yeast would be . . . I don't know, a Tesla, or a Transformer robot.

"Yeast has contributed more to human happiness than any organism on the planet, through baking, brewing and wine making," says Ian Paulsen, director of the ARC Centre of Excellence in Synthetic Biology in Sydney, Australia. "It is truly a work-horse organism. It's safe, we eat it, no problem, it's not a pathogen, and we can work it at massive scale." This last point is crucial. Yeast is one of the only organisms that doesn't have problems with viral infections, whereas bacteria, when grown in bulk, get infected.

Each group in the consortium to create synthetic yeast, which is led by Paulsen, was given a different chromosome of yeast to work on. All have now been synthesized; the job is putting them together in a cell and getting it to work. This task has proven far harder than first thought, because there are codependencies between DNA on different chromosomes, which means the engineered chromosomes don't necessarily behave the same way as the original ones when they're in the same cell. It's a difficult job, but surmountable. When synthetic yeast is finally ready, it will be proof of principle that you can make synthetic mouse or human cell lines for personalized medicine and drug development.

Paul Freemont, at Imperial College London, is one of the world's leaders in synthetic biology. He's already had the same umbrella idea we promoted for bringing together researchers in artificial intelligence, launching the Global Biofoundries Alliance in 2019. A biofoundry is a center where engineers, biologists, and geneticists are brought together to work on designing biological systems and components. At a working level right now, that means making enzymes, using genetic engineering to make cells able to produce pharmaceuticals, or indeed to function as tests for the presence of other molecules or viruses. The London biofoundry, for example, took three weeks to create a cell line able to diagnose COVID-19.

The alliance brings together some twenty-seven international institutions, including from the UK, US, Japan, Singapore, China, Australia, Denmark, and Canada, with the aim of accelerating research in synthetic biology. We should add our own investment to the Alliance to further accelerate their work, at the same time making it all open-access and shareable, and (as the Alliance is doing) devoting significant activity to assessing safety issues.

Once synthetic biologists have created the new yeast, they will have an organism that can be tweaked to do different things. The ambition is to deliver a sustainable future. As we all know, our civilization is built on a petrochemical base that is unsustainable and that has pushed us into a life-threatening climate crisis. This base must be replaced with one that is sustainable and biologically based. Synthetic yeast could break down sugars in crops and make them into fuels, but this is currently hard to do at scale because we don't have the conversion efficiency. This is something to work on. Before then, we can more easily make solvents, agricultural and industrial chemicals with a reprogramed yeast. Claudia Vickers, a synthetic biologist at Queensland University in Australia, for example, is using microbes to produce a plant hormone that can increase yields and act as a pesticide.

We will make organisms on a large scale that can sequester carbon dioxide, de-acidify the oceans, de-pollute spoiled environments, manufacture pharmaceuticals (including new antibiotics and vaccines), be used to grow plentiful and nutritious food, purify water, produce sustainable construction materials, and break down our waste.

~

IT IS THIS POTENTIAL that has proponents claiming that in this century we will start to get real control over the fundamentals of biology, and that synthetic biology is the field that will make the most profound changes to the way we live. To do this, we will need to define the set of rules—it will be a complex set, for sure—for building a living system.

Synthetic biologists are inspired by the physicist Richard Feynman, who said, "What I cannot create, I do not understand." The aim is to build modules for different aspects of cell function: for energy, metabolism, motility, sensing. We should be able to put these components together and build a variety of living systems. "To the extent that we can build a computer or very complex system with inert nonliving matter, can we apply the same approach and level of sophistication to living matter?" says Paul Freemont. "I still think we can."

There's a microbrewery just down the road from my house. There's probably one near you, too. Craft beer is a big deal in the drinks industry these days. Big tanks with yeast inside making alcohol doesn't bother anyone.

What we can do—right now—is set up craft biorefineries alongside the beer. As you buy your alcohol, you can buy some other biorefined product. Perfume, perhaps. Or food. We might have small biorefineries that recycle our organic waste. This won't require completely synthetic yeast; we can make do for now with tweaked, genetically engineered yeast. The aim is to soften up the public for having synthetic organisms as part of our lives, and to show that the concepts of synthetic biology are marketable, especially for high-value products such as fragrance.

The making of the beer itself can easily become more efficient, too. Flowers from the hops plant are used to flavor many beers, but they are expensive to grow and require huge amounts of water. Another problem is that the "hoppiness" of the flowers—a flavor deriving from the oil content—isn't constant, so it is hard to standardize the taste of the beer. Charles Denby at the University of California, Berkeley, and colleagues, engineered yeast to express genes that produce hoppy flavors and used the yeast to brew beer.[15] The Berkeley group also made genetically modified yeast able to produce THC and other cannabis compounds.[16] Traditionally a lot of cannabis has been consumed at Berkeley, but probably few stoners appreciate that cannabis is another crop that is costly to produce (by some reports accounting for a stunning 3 percent of energy usage in California). Yeast able to make cheaper, better, more ecologically sustainable beer and weed might help at least a certain sector of the public.

All living systems have the same code, and use the same machinery. By building a completely synthetic version of a living system, we will need to consider new philosophical and moral questions, but the aim is to build and harness safe systems that may not exist in the natural world.

If, starting with yeast, we can learn to build with synthetic eukaryotes, we will be on a path that might eventually allow us to build synthetic multicellular organisms. You can imagine various changes that might be seen as desirable in people living on the Moon or Mars, for example. Adaptations to lower gravity, higher radiation; cells able to synthesize nutrients and vitamins that are hard to come by off-planet. Similar approaches to redesigning life might be taken with other animals, and plants. I can imagine lichens and algae engineered with the ability to grow on the surface of Mars, for example.

Our "second genesis" approach can be summarized thus: Spend a lot of money on basic research, and make sure the work is shared and publicly available. That is the way we're going to make progress on the ultimate questions of consciousness and life, but along the way we're going to develop a lot of very useful things. Artificial intelligence that can solve the energy problem, perhaps by figuring out how to make nuclear fusion work, or getting more out of wind and solar power; engineered cells that can create the commodities that we currently rely on oil to manufacture. It's easy to forget that we don't just burn oil; we process it to make hundreds of things we use in our daily lives, from running shoes and clothing to sunglasses, lipstick, and aspirin. We need to make these things without using oil, and the tools of synthetic biology will allow us to.

I think that along the way the ultimate questions will start to dissolve. When Venter's group took a synthetic genome and loaded it into an empty bacterial cell, the process was described (I did the same just a few pages ago) in the same way we talk about rebooting a computer. That's perhaps because it feels too startling to say "life was initiated" as if "life" is a mechanical switch you can throw, Frankenstein-like. But that's what actually happened. Per-haps the significance didn't sink in because it was "only" bacteria coming to life. But this kind of work, and the kind we're funding in this chapter, will bring home how there is nothing mysterious about life. That's not to say there's nothing *wonderful* about it, nor that we shouldn't cherish and protect it. As we create new forms of life, we should take great care to protect existing life on the planet: That is the core reason we're embarking on the path laid out by synthetic biology.

As we become able to build life-forms, we will start to see that life is not something that is conveyed by a vital force, as in the veins of Hephaestus's robots. Life is extraordinary among phe-nomena that we know of, because it can resist entropy. For a time, at least, it works to avoid the second law of thermodynamics, the decay and the collapse of order, by constantly shuttling energy. As biochemist Nick Lane of University College London puts it, "If life is nothing but an electron looking for a place to rest, death is nothing but that electron come to rest." That is the basis of the wonder of life.

For consciousness, too, machine learning is going to provide more and more benefits as it develops. Again there are dangers, and again they are around security and safety. These mustn't be neglected. It might be, though, that the hand-wringing over whether an artificial system can experience the smell of coffee will diminish as our systems get better and as our empathy for algorithms increases.

Venter embedded the names of the scientists involved in the creation of the synthetic bacterium into the organism's DNA code. He also included this quote from James Joyce's *A Portrait of the Artist as a Young Man*: "To live, to err, to fall, to triumph, to re-create life out of life."

It's a great quote, to be sure. But I wish he had also inserted this line, spoken by the superbeing Dr. Manhattan in Alan Moore's

Watchmen: "A live body and a dead body contain the same number of particles. Structurally, there's no discernible difference. Life and death are unquantifiable abstracts."

Achieved

An expansion in deep-learning facilities and quantum computing research, and increased transparency about progress and aims. A better understanding of consciousness achieved by closely mimicking its processes in machines. Basic synthetic bacterial organisms made at scale for multiple purposes; progress on creating synthetic "higher" life-forms based on yeast cells. Stretch goals of two genesis events: a machine with human-level intelligence; a synthetic eukaryote life-form with functions and genome that have not evolved by natural selection.

Money spent

Research toward the creation of artificial
general intelligence: .... $100 billion

The Tangled Bank organization for
quantum computing: .... $100 billion

Creation of the Synthetic Alliance, an organization
aiming to develop artificial and synthetic life-forms: .... $100 billion

Development of craft biorefineries: .... $10 billion

Total: .... **$310 billion**

The famous "Blue Marble"—the Earth seen from Apollo 17, December 1972.

Epilogue

How to Spend It

How to Save the World for Just a Trillion Dollars started out as a bit of fun. But what began as a flight of fancy quickly started to seem real as I researched the book and considered the realities of what could—and should—be done to address such issues as global inequality, the slow pace of scientific discovery, the lack of action on climate change, and the sheer amount of money that is wasted or hoarded each year.

The ten mega projects in this book are all things that the world could do. The resources and, in most cases, the technical knowhow is there. Of course the political and social will is not, but I can't get over the potential that we *have* to do these things. The frustrated potential.

If this was a trillion-dollar version of *Brewster's Millions*, where I had to pick one of my ten options, what would I do? I've become attached to all the projects. Many of the scientists I've spoken to have made convincing and impassioned cases, and I can see the benefits of them all.

It's dismaying that some of the big science questions we've looked at—amounting to some of the biggest questions humans have ever asked—hardly require much money, and yet researchers are left to scratch around for funding. As a fellow scientist I'm sorry not to help out.

The argument for spending the money on universal education and cash transfers is almost impossible to resist. These are people's lives, right now. Money spent on education programs would change the lives of hundreds of millions of children forever. And, faced with real lives, it's hard to justify spending money on physics and understanding the nature of reality, or on the attempt to create an artificial intelligence with human-level ability, or for starting a permanent Moon base. We could change the lives of millions of people beyond recognition and it seems criminal not to do so. So the biggest ethical problem I've had—and it has genuinely weighed upon me—is in deciding *not* to give most of the money to the world's poorest people.

For with this much money we have a chance to change the world—and perhaps save it—by tackling climate change with everything we've got. If we *don't* do this, starting right now, the future for many of the world's poorest people will be far worse than their present. So the $1 trillion, inevitably, has to go toward tackling the climate crisis. And here's what I would do. Give $500 billion to projects to transition the world to renewable energy (chapter 3), and $500 billion to a massive biodiversity renewal and carbon drawdown scheme (chapters 4 and 7).

First, renewable energy. I am quite optimistic about the growth and uptake of wind and solar energy. Prices for renewable energy are falling all the time, and many countries, as well as eleven US states, have made legally binding commitments to reach net zero by 2050 (or, in China's case, 2060). But the transition does need a push. We need to cut carbon emissions by at least 45 percent by 2030 to keep global warming to "manageable" levels. Investment is needed in the Global south, and especially in India, which has the fastest-growing power demand of any country. Most of India's energy (around 75 percent) is still supplied by coal, and the prime minister, Narendra Modi, wants to auction forty-one coal mines to private companies to speed their development. I'd put money into India to get it off coal faster, to install wind and solar power, and to modernize its electricity distribution grid and storage capability for renewable energy. Thus, India gets $200 billion; renewables projects in the United States and China get $100 billion each. Another $100 billion goes on pushing renewable energy in the rest of the world.

And so to the other half trillion.

It's going to take some years, probably decades, to get to 100 percent renewable energy, to get to net-zero carbon emissions. The transition is too slow and, even when we're there, that alone is not going to protect our future. We've seen how most of the planetary safety boundaries have been broken, and how the extinction crisis is threatening the collapse of global ecosystems. Thankfully, there is a way to make time to decarbonize our civilization and to safeguard biodiversity: Buy up and protect areas of key importance for biodiversity, for carbon storage, and for carbon drawdown. There are more than two million square miles around the world that could simply be set aside, protected from grazing or development, and allowed to regrow, leading to the capture, over the next three decades, of tens of billions of tons of carbon dioxide. There is more land, too, that urgently needs proper protection.

As a very rough estimate, we saw that purchasing, managing, and protecting land for all endangered species, and factoring in support for people currently living in these areas, would cost about $100 billion per year. Most of our $500 billion, therefore, should go into a program of ecosystem renewal and protection. It will include a mixture of mass tree-planting and regrowth and restoration projects, including of marine and freshwater systems. Some of this spending will mean large areas of land currently used for cattle farming and for growing food for cattle is purchased and allowed to regenerate naturally. It is my hope and expectation that this drives down the consumption of beef globally and stimulates the development of plant-based alternatives.

It's important to state that half a trillion dollars spent on ecosystem renewal won't save us from climate change. But, if we get it right, simply letting forests grow is a powerful method for capturing carbon and increasing biodiversity and giving us time to get the rest of our society decarbonized. We do, of course, have to make sure the forests are not cleared as soon as they've grown—perhaps payments for ecosystem services could be introduced here.

I would also like to save a small amount for other projects. $1 billion for trials of enhanced weathering (chapter 7), which seems to have great potential for capturing carbon dioxide, and $19 billion to the African Space Agency to establish the Terran Alliance for the Moon (chapter 5). NASA is attempting to control commercial activities on the Moon and other celestial bodies with its Artemis Accords.[1] It's vital that other countries have a stake in the development of lunar missions and a say in space regulation; this investment will help ensure that. As a bonus, and a slight cheat, we could attach conditions to the money. In other words, tie the biodiversity restoration schemes, and the spending on the development and installation of renewables in the Global South, to education and poverty reduction programs, so as to address directly the humanitarian concerns of chapter 1.

That's it. That's all the money spent.

Now it's time to drift back to the real world.

∽

IN HIS FIRST YEAR IN OFFICE, PRESIDENT JOE BIDEN tabled a $6 trillion spending plan, with around $2 trillion earmarked for climate-related projects. The European Union has also pledged big on climate, passing a €1.074 trillion budget for 2021 to 2027, with €550 billion of that directly aimed at "green" projects and the rest tagged with environmental "do no harm" clauses. Well, it's a start. The EU needs an estimated €2.4 trillion in low-carbon investments by 2027 to meet its own emissions targets.[2]

The UN has seventeen Global Goals that overlap with many of the projects in this book (but don't include science or space exploration). Officially called the Sustainable Development Goals, they are templates that, if implemented, will lead to (for example) an end to poverty, universal education, clean energy, action on climate change. The goals have been agreed upon by all 193 UN member states, but have no legal power. The UN estimates it would cost $175 billion per year to eliminate extreme poverty around the world. To achieve all seventeen goals, including action on global health, climate change, biodiversity, sustainable housing, industry, and food production and consumption, the UN puts the figure at $2 trillion—that's 2 percent of world GDP—per year. The UN itself can't get close. Its budget for 2020 was a pitiful $3.1 billion.[3] But, as we've seen, it's not an impossible amount of money in our brave new post-COVID world.

In 2020, globally, somewhere between $9 trillion and $12 trillion was found or created—and spent—in response to the coronavirus crisis. Clearly, we need to "find" more for the climate and biodiversity emergency, and for UN Global Goals. Each year $1 trillion is spirited out of the economies of the Global South by multinationals shifting money offshore to avoid tax; trillions more goes missing because people in the Global South working for export industries are underpaid.[4] If you were an omnipotent world ruler, if you could snap your fingers like an ecologically friendly Thanos and redistribute global wealth, you might be able to rustle up the money. Until then, we have to apply political pressure, agitate, and vote with our wallets.

We have to become more aware of what our consumption is doing. Remember that each year we take resources from the planet, stuff to eat, to burn for energy, to build with or turn into some product. In 2017, the total weight of all the stuff we take came in at 101 billion tons. The maximum sustainable amount we can take each year is 55 billion tons.[5] In our overconsumption we are sawing off the branch we are sitting on. It simply can't go on, and when I say "It can't go on," I don't mean it in the desk-thumping way a schoolteacher once told me, I mean in a literal sense: It *can't* go on, because the system will collapse.

It's easy to despair and it's normal to feel anxious. I've done both. What I have found warming, and sustaining, is talking to some the thousands of scientists and economists and activists around the world who are working for a better world. There are millions of people in this team, in this project to change the world. They are devoting their lives to it. The novelist and political activist Arundhati Roy said that the coronavirus pandemic is a portal, a gateway between one world and the next.[6] We are choosing at the moment to step through the portal into a world where ecosystem collapse is a likelihood. We have to choose, now, to make a different world.

Global wealth is greater by far than at any point in human history. Philanthropy is growing. Let's hope it grows more, and that nation states act together, across borders and with private enterprise. Personally, I'll never have anything like the sums talked about in this book, but I have something. I am happy for my taxes to be used in the great transition and redistribution. And I am happy to try to ensure that my own spending, and whoever invests my pension, favors goods and services with low-carbon, carbon-neutral, or even carbon-negative footprints. Governments have to change and so do the rest of us.

No one is alone in this. The hope that we can genuinely step through the portal into a post-Covid society that is greener and more equitable—the desire to do so—is shared by a majority of people. There is a togetherness here, and there is succor from the knowledge that we are part of the most important collective mission of all time.

Notes

INTRODUCTION

1. Elizabeth Schulze, "The Fed launched QE nine years ago—these four charts show its impact," CNBC, November 24, 2017.

2. "Global Wealth Report 2019," Credit Suisse, credit-suisse.com.

3. Kate Rooney, "Private equity's record $1.5 trillion cash pile comes with a new set of challenges," CNBC, January 3, 2020.

4. "How to Grow Green," Bloomberg Green, June 9, 2020.

5. "Global QE asset purchases to reach USD6 trillion in 2020," Fitch Ratings, April 24, 2020.

CHAPTER 1

1. "Economists are rethinking the numbers on inequality," *The Economist*, November 28, 2019.

2. "COVID-19 (social protection): statements," Tithe an Oireachtais/Houses of the Oireachtas, April 2, 2020, oireachtas.ie/en.

3. Anthony Leonardi, "'Take dramatic action': AOC calls for universal basic income as response to coronavirus," *Washington Examiner*, March 12, 2020.

4. Sara Clarke, "States with the most billionaires," *U.S. News,* March 8, 2019.

5. "Mukesh Ambani's Reliance Industries becomes world's 6th largest oil company," *India Today,* November 20, 2019.

6. David K. Evans and Anna Popova, "Cash transfers and temptation goods: a review of global evidence," World Bank Group, Impact Evaluation series no. IE 127, Policy Research working paper no. WPS 6886, May 1, 2014.

7. Ibid.

8. Oriana Bandiera, et al., "Labor Markets and Poverty in Village Economies," *Quarterly Journal of Economics* 2, no. 132 (March 20, 2017): 811–70.

9. Abhijit Banerjee, et al., "A multifaceted program causes lasting progress for the very poor: Evidence from six countries," *Science* 6236, no. 348 (May 15, 2015).

10. Suresh de Mel, et al., "One-Time Transfers of Cash or Capital Have Long-Lasting Effects on Microenterprises in Sri Lanka," *Science* 6071, no. 335 (February 24, 2012): 962–66.

11. Christopher Blattman, et al., "The Returns to Microenterprise Support among the Ultrapoor: A Field Experiment in Postwar Uganda," *American Economic Journal: Applied Economics* 2, no. 8 (April 2016): 35–64.

12. Dennis Egger, et al., "General Equilibrium Effects of Cash Transfers: Experimental Evidence from Kenya," *National Bureau of Economic Research,* December 2019.

13. Damon Jones and Ioana Marinescu, "The Labor Market Impacts of Universal and Permanent Cash Transfers: Evidence from the Alaska Permanent Fund," *National Bureau of Economic Research,* February 2018.

14. Jesse Cunha, et al., "The Price Effects of Cash Versus In-Kind Transfers," *Review of Economic Studies* 1, no. 86 (January 2019): 240–81.

15. Laurence Chandy, et al., *The Last Mile in Ending Extreme Poverty* (Brookings Institution Press, 2015).

16. Decca Aitkenhead, "Abhijit Banerjee: 'The poor, probably rightly, see their chances of getting somewhere different are minimal,'" *The Guardian*, April 22, 2012.

17. Abhijit Banerjee, et al., "Effects of a Universal Basic Income during the pandemic," *Innovations for Poverty*, September 2, 2020.

18. Stephen Kidd, "The demise of Mexico's Prospera programme: a tragedy foretold," Development Pathways, June 2, 2019.

19. Najy Benhassine, et al., "Turning a shove into a nudge? A 'labeled cash transfer' for education." *National Bureau of Economic Research*, July 2013.

20. Charles Murray, *In Our Hands: A Plan to Replace the Welfare State* (American Enterprise Institute Press, 2016).

21. Richard Akresh, et al., "Long-term and Intergenerational Effects of Education: Evidence from School Construction in Indonesia," *National Bureau of Economic Research*, November 2018.

22. Rebecca Winthrop and Homi Kharas, "Want to save the planet? Invest in girls' education," Brookings, March 3, 2016.

23. "Sector summary: health and education," Project Drawdown, drawdown.org.

CHAPTER 2

1. Patrick Walker, et al., "Report 12—The global impact of COVID-19 and strategies for mitigation and suppression," *Science* 6502, no. 369 (December 6, 2020).

2. "Cardiovascular diseases (CVDs)," World Health Organization (WHO), June 11, 2021, who.int; William M. Carroll, "The global burden of neurological disorders,"

The Lancet Neurology 5, no. 18 (March 14, 2019); "Cancer," WHO, September 21, 2021, who.int; "Infectious diseases kill over 17 million people a year: WHO warns of global crisis," World Health Organization, January 1, 1996, who.int.

3. Matthew Young, et al., "Single-cell transcriptomes from human kidneys reveal the cellular identity of renal tumors," *Science* 6402, no. 361 (August 10, 2018): 594–99.

4. Lindsey W. Plasschaert, et al., "A single-cell atlas of the airway epithelium reveals the CFTR-rich pulmonary ionocyte," *Nature* 7718, no. 560 (August 1, 2018): 377–81.

5. Ceri Parker, "What if we get things right? Visions for 2030," World Economic Forum, October 29, 2019, weforum.org.

6. Dean T. Jamison, et al., "Global health 2035: A world converging within a generation," *The Lancet* 9908, no. 382 (December 3, 2013): 1898–1955.

7. "Cumulative and confirmed and probable COVID-19 cases reported by Countries and Territories in the Region of the Americas," Pan American Health Organization, 2020, ais. paho.org.

8. "Ethiopia," WHO, who.int; "Ghana," WHO, afro.who.int.

9. Shreeshant Prabhakaran, et al., "Financial sustainability of Indonesia's Jaminan Kesehatan Nasional: Performance, Prospects, and Policy Options," Health Policy Plus and Tim Nasional Percepatan Penanggulangan Kemiskinan (TNP2K), May 2019.

10. Emmanuel Akinwotu, "Africa declared free of wild polio after decades of work," *The Guardian*, August 25, 2020.

11. "Poliomyelitis," WHO, July 22, 2019, who.int.

12. "Ten threats to global health in 2019," WHO, 2019; "Thirteenth General Programme of Work 2019–2023," WHO, 2019.

13. "The urgency of now," The Coalition for Epidemic Preparedness Innovations, endpandemics.cepi.net.

14. Michael Mina, et al., "Science Forum: A Global Immunological Observatory to meet a time of pandemics," *eLife*, June 8, 2020.

15. Michael Le Page, "First evidence that GM mosquitoes reduce disease," *New Scientist*, July 15, 2016.

16. C. M. Collins, et al., "Effects of removal or reduced density of the malaria mosquito, *Anopheles gambiae s.l.*, on interacting predators and competitors in local ecosystems," *Medical and Veterinary Entomology* 1, no. 33 (July 25, 2018): 1–15.

17. Max Roser, et al., "Life Expectancy," Our World in Data, October 2019, ourworldindata.org.

18. Vasilis Kontis, et al., "Future life expectancy in 35 industrialised countries: projections with a Bayesian model ensemble," *The Lancet* 10076, no. 389 (February 21, 2017): 1323–35.

19. Francine E. Garrett-Bakelman, et al., "The NASA Twins Study: A multidimensional analysis of a year-long human spaceflight," *Science* 6436, no. 364 (April 12, 2019).

20. Chelsea Gohd, "Can We Genetically Engineer Humans to Survive Missions to Mars?," Space.com, November 7, 2019.

21. David Cyranoski, "Russian 'CRISPR-baby' scientist has started editing genes in human eggs with goal of altering deaf gene," *Nature*, October 18, 2019.

22. Giorgio Sirugo, et al., "The Missing Diversity in Human Genetic Studies," *Cell* 1, no. 177 (March 21, 2019): 26–31.

23. Ergin Beyret, et al., "Single-dose CRISPR-Cas9 therapy exceeds lifespan of mice with Hutchinson–Gilford progeria syndrome," *Nature Medicine* 25 (February 18, 2019): 419–22.

24. Cori Bargmann, "How the Chan Zuckerberg Science Initiative plans to solve disease by 2100," *Nature*, January 3, 2018.

25. "Global death rate from rising temperatures projected to surpass the current death rate of all infectious diseases combined," Climate Impact Lab, August 3, 2020, impactlab.org.

CHAPTER 3

1. Raymond P. Sorenson, "Eunice Foote's Pioneering Research On CO_2 and Climate Warming," *Search and Discovery*, January 31, 2011.

2. Piers M. Forster, et al., "Current and future global climate impacts resulting from COVID-19," *Nature Climate Change* 10 (August 7, 2020): 913–19.

3. Ove Hoegh-Guldberg, et al., "The human imperative of stabilizing global climate change at 1.5°C," *Science* 6459, no. 365 (September 20, 2019).

4. Solomon Hsiang, et al., "Estimating economic damage from climate change in the United States," *Science* 6345, no. 356 (June 30, 2017): 1362–69.

5. Fiona Harvey, "UK facing worst wheat harvest since 1980s, says farmers' union," *The Guardian*, August 17, 2020.

6. Agence France-Presse in Stockholm, "'The climate doesn't need awards': Greta Thunberg declines environmental prize," *The Guardian*, October 29, 2019.

7. Jonathan Watts, et al., "Oil firms to pour extra 7m barrels per day into markets, data shows," *The Guardian*, October 10, 2020.

8. "China Wants to Be Carbon Neutral By 2060. Is That Possible?" Bloomberg Green, September 23, 2020.

9. Nicholas Stern and Simon Dietz, "Growth and Sustainability: 10 years on from the Stern Review," London School of Economics and Political Science, October 27, 2016.

10. Mark Jacobson, et al., "100% Clean and Renewable Wind, Water, and Sunlight All-Sector Energy Roadmaps for 139 Countries of the World." *Joule* 1, no. 1 (September 6, 2017): 108–21.

11. V. Masson-Delmotte, et al., "Global Warming of 1.5°C," Intergovernmental Panel on Climate Change, 2018.

12. D. L. Elliott, et al., "An Assessment of the Available Windy Land Area and Wind Energy Potential in the Contiguous United States," Pacific Northwest Laboratory, 1991.

13. Jack Unwin, "AWEA names Tri Global Energy leading wind developer in Texas," Power Technology, February 6, 2019.

14. Zhenzhong Zeng et al., "A reversal in global terrestrial stilling and its implications for wind energy production," Nature Climate Change 9 (November 18, 2019): 979–85.

15. Adam Vaughan, "Mersey feat: world's biggest wind turbines go online near Liverpool," The Guardian, May 17, 2017.

16. Anne Bergen, et al., "Design and in-field testing of the world's first ReBCO rotor for a 3.6 MW wind generator," Superconductor Science and Technology 12, no. 32 (October 25, 2019).

17. "Kenya launches Africa's largest wind farm," Africa News, July 20, 2019.

18. "Frequently asked questions (FAQs): How much does it cost to generate electricity with different types of power plants?," U.S. Energy Information Administration, eia.gov.

19. "Lazard's levelized cost of energy analysis—version 12.0," Lazard, November 2018.

20. José Rojo Martín, "MoU signed for 2.6GW Mecca solar programme," PVTech, March 25, 2019.

21. Charlotte Vogt, et al., "The renaissance of the Sabatier reaction and its applications on Earth and in space," Nature Catalysis 2 (March 11, 2019): 188–97.

22. Chris Goodall, What We Need to Do Now: For a Zero Carbon Future (Profile Books, 2020).

23. Gunther Glenk and Stefan Reichelstein, "Economics of converting renewable power to hydrogen," Nature Energy 4, (February 25, 2019): 216–22.

24. "'Hydrogen Economy' Offers Promising Path to Decarbonization," BloombergNEF, March 30, 2020.

25. Peter Behr, "Details emerge about DOE 'super-grid' renewable study," *E&E News*, October 30, 2019.

26. Peter Fairley, "How a Plan to Save the Power System Disappeared," *The Atlantic*, August 20, 2020.

27. Mark Z. Jacobson, et al., "100% Clean and Renewable Wind, Water, and Sunlight All-Sector Energy Roadmaps for 139 Countries of the World," *Joule* 1, no. 1 (September 6, 2017): 108–21.

28. Mark Z. Jacobson, et al., "Matching demand with supply at low cost in 139 countries among 20 world regions with 100% intermittent wind, water, and sunlight (WWS) for all purposes," *Renewable Energy* 123 (August 2018): 236–48.

29. Emil G. Dimanchev, et al., "Health co-benefits of sub-national renewable energy policy in the US," *Environmental Research Letters* 8, no. 14 (2019).

30. Mark Z. Jacobson et al., "Impacts of Green New Deal Energy Plans on Grid Stability, Costs, Jobs, Health, and Climate in 143 Countries," *One Earth* 4, no. 1 (December 2019): 449–63.

31. "Fort Calhoun becomes fifth U.S. nuclear plant to retire in past five years," U.S. Energy Information Administration, October 31, 2016, eia.gov.

32. Jillian Ambrose, "New windfarms will not cost billpayers after subsidies hit record low," *The Guardian*, September 20, 2019.

33. "Small modular reactors," International Atomic Energy Agency, iaea.org.

34. "About Us," NuScale Power, nuscalepower.com.

35. International Atomic Energy Agency, op. cit.

36. "Science and Technology Strategy," National Nuclear Laboratory, nnl.co.uk.

37. Mariliis Lehtveer, et al., "What Future for Electrofuels in Transport? Analysis of Cost Competitiveness in Global Climate Mitigation," *Environmental Science & Technology* 3, no. 53 (February 5, 2019): 1690–97.

38. Ilkka Hannula and David M. Reiner, "Near-term potential of biofuels, electrofuels, and battery electric vehicles in decarbonizing road transport," *Joule* 10, no. 3 (October 16, 2019): 2390–2402.

39. GL Reynolds, "The multi-issue mitigation potential of reducing ship speeds," Seas at Risk, November 11, 2019.

40. Jacob Mason, et al., "A Global High Shift Cycling Scenario: The Potential for Dramatically Increasing Bicycle and E-bike Use in Cities Around the World, with Estimated Energy, CO_2, and Cost Impacts," Institute for Transportation & Development Policy, November 12, 2015.

41. David Coady, et al., "Global Fossil Fuel Subsidies Remain Large: An Update Based on Country-Level Estimates," IMF Working Paper, May 2019.

CHAPTER 4: SAVE LIFE ON EARTH

1. Stuart L. Pimm, et al., "The biodiversity of species and their rates of extinction, distribution, and protection," *Science* 6187, no. 344 (May 2014).

2. Malcolm L. McCallum, "Amphibian Decline or Extinction? Current Declines Dwarf Background Extinction Rate," *Journal of Herpetology* 3, no. 41 (September 2007): 483–91.

3. Anthony D. Barnosky, et al., "Has the Earth's sixth mass extinction already arrived?" *Nature* 471 (March 3, 2011): 51–57.

4. Jun-xuan Fan, et al., "A high-resolution summary of Cambrian to Early Triassic marine invertebrate biodiversity," *Science* 6475, no. 367 (January 17, 2020): 272–77.

5. Alejandro Estrada, et al., "Impending extinction crisis of the world's primates: Why primates matter," *Science Advances* 1, no. 3 (January 18, 2017).

6. Francisco Sánchez-Bayo and Kris A. G. Wyckhuys, "Worldwide decline of the entomofauna: a review of its drivers," *Biological Conservation* 232 (April 2019): 8–27.

7. Rodolfo Dirzo, et al., "Defaunation in the Anthropocene," *Science* 6195, no. 345 (July 25, 2014): 401–6.

8. Charles Piller, "Verily, I swear. Google Life Sciences debuts a new name," *Stat News*, December 7, 2015.

9. Sandra Díaz, et al., "Assessing nature's contributions to people," *Science* 6373, no. 359 (January 19, 2018): 270–72.

10. David Tilman, et al., "Biodiversity and ecosystem functioning," *Annual Review of Ecology, Evolution, and Systematics* 45 (October 1, 2014): 471–93.

11. Robert Costanza, et al., "Changes in the global value of ecosystem services," *Global Environmental Change* 26 (May 2014): 152–58.

12. John Harte and Rebecca Shaw, "Shifting Dominance Within a Montane Vegetation Community: Results of a Climate-Warming Experiment," *Science* 5199, no 267 (February 10, 1995): 876–80.

13. Laura E. Koteen, et al., "Invasion of non-native grasses causes a drop in soil carbon storage in California grasslands," *Environmental Research Letters* 4, no. 6 (October 10, 2011).

14. James E. M. Watson, et al., "The exceptional value of intact forest ecosystems," *Nature Ecology & Evolution* 2 (February 26, 2018): 599–610.

15. Adam Vaughan, "Amazon deforestation officially hits highest level in a decade," *New Scientist*, November 18, 2019.

16. Thomas E. Lovejoy and Carlos Nobre, "Amazon Tipping Point," *Science Advances* 2, no. 4 (February 21, 2018).

17. Susan C. Cook-Patton, et al., "Mapping carbon accumulation potential from global natural forest regrowth," *Nature* 585, (September 23, 2020): 545–50.

18. Jon Lee Anderson, "Letter from the Amazon," *The New Yorker*, November 4, 2019.

19. Anna Gross and Andres Schipani, "Brazil tells rich countries to pay up to protect Amazon," *Financial Times*, October 7, 2019.

20. Ove Hoegh-Guldberg, "Coral reefs: megadiversity meets unprecedented environmental change," *Biodiversity and Climate Change: Transforming the Biosphere*, eds. Thomas Lovejoy and Lee Hannah (Yale University Press, 2019).

21. K. Frieler, et al., "Limiting global warming to 2 degrees C is unlikely to save most coral reefs," *Nature Climate Change* 2, no. 3 (September 16, 2012): 165–70.

22. Dylan Chivian, et al., "Environmental Genomics Reveals a Single-Species Ecosystem Deep Within Earth," *Science* 5899, no. 322 (October 10, 2008): 275–78.

23. Camilo Mora, et al., "How Many Species Are There on Earth and in the Ocean?" *PLOS Biology*, August 23, 2011.

24. Kenneth J. Locey and Jay T. Lennon, "Scaling laws predict global microbial diversity," *PNAS* 21, no. 113 (March 30, 2016): 5970–75.

25. Alexander Nater, et al., "Morphometric, Behavioral, and Genomic Evidence for a New Orangutan Species." *Current Biology* 2, no. 27 (November 20, 2017): 3487–98.

26. Isabelle B. Laumer, et al., "Orangutans (*Pongo abelii*) make flexible decisions relative to reward quality and tool functionality in a multi-dimensional tool-use task," *PLOS ONE*, February 13, 2019.

27. Erik Stokstad, "New great ape species found, sparking fears for its survival," *Science*, November 2, 2017.

28. "Wild places: Keo Seima Wildlife Sanctuary," Wildlife Conservation Society Cambodia, cambodia.wcs.org.

29. Mengey Eng, "Wildlife conservationists encouraged by Cambodia's pursuit of justice in murder case of three rangers and committed to the protection of Keo Seima Wildlife Sanctuary," Wildlife Conservation Society Cambodia, cambodia.wcs.org.

30. Robert Wallace, et al., "On a New Species of Titi Monkey, Genus *Callicebus* Thomas (Primates, *Pitheciidae*), from Western Bolivia with Preliminary Notes on Distribution and Abundance," *Primate Conservation* 20 (May 2006): 29–39.

31. Donal McCarthy, et al., "Financial Costs of Meeting Global Biodiversity Conservation Targets: Current Spending and Unmet Needs," *Science* 6109, no. 338 (October 11, 2012): 946–49.

32. "Key Biodiversity Areas: keep nature thriving," Key Biodiversity Areas, keybiodiversityareas.org.

33. Michael Soulé and Reed Noss, "Rewilding and Biodiversity: Complementary Goals for Continental Conservation," *Wild Earth* 3, no. 8 (1998): 19–28.

34. Edward O. Wilson, *Half-Earth: Our Planet's Fight for Life* (W. W. Norton, 2016).

35. Jonathan Baillie and Ya-Ping Zhang, "Space for nature," *Science* 6407, no. 361 (September 14, 2018): 1051.

36. Patrick Barkham, "First wild stork chicks to hatch in UK in centuries poised to emerge," *The Guardian*, April 26, 2020.

37. Eric Dinerstein, et al., "A 'Global Safety Net' to reverse biodiversity loss and stabilize Earth's climate," *Science Advances* 36, no. 6 (September 4, 2020).

38. "United to Reverse Biodiversity Loss by 2030 for Sustainable Development," Leaders Pledge for Nature, leaderspledgefornature.org.

39. Mark Bush, "A neotropical perspective on past human–climate interactions and biodiversity," *Biodiversity and Climate Change*, eds. Thomas Lovejoy and Lee Hannah (Yale University Press, 2019).

40. Céline Bellard, et al., "Alien species as a driver of recent extinctions," *Biology Letters* 2, no. 12 (February 2016).

41. *Global Biodiversity Outlook 5*, Convention on Biological Diversity, August 18, 2020.

42. Sandra Díaz, et al., "Pervasive human-driven decline of life on Earth points to the need for transformative change," *Science* 6471, no. 366 (December 13, 2019).

CHAPTER 5: SETTLE OFF-PLANET

43. "Humanity is returning to the Moon. How we do it matters," Open Lunar Foundation, openlunar.org.

44. Oliver Morton, *The Moon: A History for the Future* (Profile Books, 2019).

45. Scott Kelly, *Endurance: The Extraordinary True Story of My Year in Space* (Penguin, 2017).

CHAPTER 6: FIND SOME ALIENS

1. Guy Consolmagno, "Dreaming Martians," *L'Osservatore Romano*, 2018.

2. Abby Ohlheiser, "Why the Vatican doesn't think we'll ever meet an alien Jesus," *Washington Post*, August 1, 2015.

3. David Willey, "Vatican says aliens could exist," *BBC*, May 13, 2008.

4. Brent Christner, et al., "Ubiquity of Biological Ice Nucleators in Snowfall," *Science* 5867, no. 319 (February 29, 2008): 1214.

5. Jane Greaves, et al., "Phosphine gas in the cloud decks of Venus," *Nature Astronomy* 5, (September 14, 2020): 655–64.

6. Jonathan Amos, "Venus: Will private firms win the race to the fiery planet?," *BBC*, September 14, 2020.

7. Dylan Chivian, et al., "Environmental Genomics Reveals a Single-Species Ecosystem Deep Within Earth," *Science* 5899, no. 322 (October 10, 2008): 275–78.

8. C. Magnabosco, et al., "The biomass and biodiversity of the continental subsurface," *Nature Geoscience* 11 (September 24, 2018): 707–17.

9. "Steven D'Hondt," University of Rhode Island, web.uri.edu.

10. J. A. Bradley, et al., "Widespread energy limitation to life in global subseafloor sediments," *Science Advances* 32, no. 6 (August 5, 2020).

11. Melissa Guzman, et al., "Identification of Chlorobenzene in the Viking Gas Chromatograph-Mass Spectrometer Data Sets: Reanalysis of Viking Mission Data Consistent with Aromatic Organic Compounds on Mars," *Journal of Geophysical Research: Planets* 7, no. 123 (June 20, 2018).

12. Gilbert V. Levin, "I'm Convinced We Found Evidence of Life on Mars in the 1970s," *Scientific American*, October 10, 2019.

13. "NASA Space Science Data Coordinated Archive," NASA, nssdc.gsfc.nasa.gov.

14. "The ExoMars Rover Instrument Suite," European Space Agency, exploration.esa.int.

15. Fred Goesmann, et al., "The Mars Organic Molecule Analyzer (MOMA) Instrument: Characterization of Organic Material in Martian Sediments," *Astrobiology* 17 (July 1, 2017): 655–85.

16. Anna Grau Galofre, et al., "Valley formation on early Mars by subglacial and fluvial erosion," *Nature Geoscience* 13 (August 3, 2020): 663–68.

17. T. Nordheim, et al., "Preservation of potential biosignatures in the shallow subsurface of Europa," *Nature Astronomy* 2 (July 23, 2018): 673–79.

18. "Life Detection Ladder," NASA, astrobiology.nasa.gov.

19. Marc Neveu, et al., "The Ladder of Life Detection," *Astrobiology* 11, no. 18 (June 4, 2018): 1375–1402.

20. M. Mastrogiuseppe, et al., "Deep and methane-rich lakes on Titan," *Nature Astronomy* 3 (April 15, 2019): 535–42; Shannon M. MacKenzie, et al., "The case for seasonal surface changes at Titan's lake district," *Nature Astronomy* 3 (April 15, 2019): 506–10.

21. Steve Oleson, "Titan Submarine—an Extraterrestrial Submarine for Titan's Cryogenic Seas," NTRS NASA, September 12, 2019, ntrs.nasa.gov.

22. Ruth-Sophie Taubner, et al., "Biological methane production under putative Enceladus-like conditions," *Nature Communications* 748, no. 9 (February 27, 2018).

23. N. Khawaja, et al., "Low-mass nitrogen-, oxygen-bearing, and aromatic compounds in Enceladean ice grains," *Monthly Notices of the Royal Astronomical Society* 4, no. 489 (October 2, 2019): 5231–43.

24. Devin Powell, "The Drake Equation Revisited: Interview with Planet Hunter Sara Seager," Space.com, September 4, 2013.

25. Anders Sandberg, et al., "Dissolving the Fermi Paradox," submitted to *Proceedings of the Royal Society of London A*, June 6, 2018.

CHAPTER 7: REDESIGN OUR PLANET

1. Donald Olson, et al., "The blood-red sky of *The Scream*," *APS News* 5, no. 13 (May 2004).

2. Fiona Harvey, "UN climate talks failing to address urgency of crisis, says top scientist," *The Guardian*, December 8, 2019.

3. Ian Joughin, et al., "Marine Ice Sheet Collapse Potentially Under Way for the Thwaites Glacier Basin, West Antarctica," *Science* 6185, no. 344 (May 16, 2014): 735–38.

4. Michaela D. King, et al., "Dynamic ice loss from the Greenland Ice Sheet driven by sustained glacier retreat," *Nature Communications Earth & Environment* 1, no. 1 (August 13, 2020).

5. Hans Joachim Schellnhuber, et al., "Why the right climate target was agreed in Paris," *Nature Climate Change* 6 (June 23, 2016): 649–53.

6. "Ecological Threat Register 2020: Understanding Ecological Threats, Resilience, and Peace," Institute for Economics & Peace, September 2020, visionofhumanity.org/indexes/ ecological-threat-register.

7. K. E. McCusker, et al., "Inability of stratospheric sulfate aerosol injections to preserve the West Antarctic Ice Sheet," *Geophysical Research Letters* (June 4, 2015).

8. Dan M. Kahan, et al., "Geoengineering and Climate Change Polarization: Testing a Two-Channel Model of Science Communication," *ANNALS of the American Academy of Political and Social Science* 1, no. 658 (February 8, 2015): 192–222.

9. Dabang Jiang et al., "Climate Change of 4°C Global Warming above Pre-industrial Levels," *Advances in Atmospheric Sciences* 35 (May 28, 2018): 757–70.

10. Jeff Tollefson, "Iron-dumping ocean experiment sparks controversy," *Nature* 7655, no. 545 (May 25, 2017): 393–94.

11. Emily Pontecorvo, "The climate policy milestone that was buried in the 2020 budget," *Grist*, January 8, 2020.

12. Jonathan Proctor, et al., "Estimating global agricultural effects of geoengineering using volcanic eruptions," *Nature* 560 (August 8, 2018): 480–83.

13. Wake Smith and Gernot Wagner, "Stratospheric aerosol injection tactics and costs in the first 15 years of deployment," *Environmental Research Letters* 12, no. 13 (November 23, 2018).

14. John C. Moore, et al., "Geoengineer polar glaciers to slow sea-level rise," *Nature*, March 14, 2018.

15. John Latham, et al., "Marine cloud brightening: regional applications," *Philosophical Transactions of the Royal Society A*, December 28, 2014.

16. Stephen Salter, et al., "Sea-going hardware for the cloud albedo method of reversing global warming," *Philosophical Transactions of the Royal Society A*, August 29, 2008.

17. Sara Budinis, et al., "An assessment of CCS costs, barriers and potential," *Energy Strategy Reviews* 22 (November 2018): 61–81.

18. Dieter Helm, *Net Zero: How We Stop Causing Climate Change* (William Collins, 2020).

19. Roelof D. Schuiling and Poppe L. de Boer, "Six commercially viable ways to remove CO_2 from the atmosphere and/or reduce CO_2 emissions," *Environmental Sciences Europe* 35, no. 25 (December 21, 2013).

20. David J. Beerling, et al., "Potential for large-scale CO_2 removal via enhanced rock weathering with croplands," *Nature* 583 (July 8, 2020): 242–48.

21. Glen Peters, "Does the carbon budget mean the end of fossil fuels?" Center for International Climate Research, April 6, 2017.

22. Sabine Fuss, et al., "Negative emissions—Part 2: Costs, potentials and side effects," *Environmental Research Letters* 6, no. 13 (May 22, 2018).

23. Rob B. Jackson, et al., "Methane removal and atmospheric restoration," *Nature Sustainability* 2 (May 20, 2019): 436–38.

24. "The Global Carbon Project," Global Carbon Project, globalcarbonproject.org.

25. Jean-François Bastin, et al., "The global tree restoration potential," *Science* 6448, no. 365 (July 5, 2019): 76–79.

26. "EU Biodiversity Strategy for 2030: Bringing nature back into our lives," European Commission, May 20, 2020.

27. Alexandra E. Petri, "China's 'Great Green Wall' Fights Expanding Desert," *National Geographic*, April 21, 2017.

28. Toby A. Gardner, et al., "A framework for integrating biodiversity concerns into national REDD+ programs," *Biological Conservation* 154 (October 2012): 61–71.

29. "Ghana deploys nature-based solutions to tackle climate change," *Ghana News Agency*, December 14, 2019.

30. Charles Eisenstein, *Climate: A New Story* (North Atlantic Books, 2018).

31. Dorte Krause-Jensen and Carlos Duarte, "Substantial role of macroalgae in marine carbon sequestration," *Nature Geoscience* 9 (September 12, 2016): 737–42.

32. Antoine de Ramon N'Yeurt, et al., "Negative carbon via Ocean Afforestation," *Process Safety and Environmental Protection* 6, no. 90 (November 2012): 467–74.

33. Thomas Wernberg, et al., "Climate-driven regime shift of a temperate marine ecosystem," *Science* 6295, no. 353 (July 8, 2016): 169–72.

34. Andrew J. Pershing, et al., "The Impact of Whaling on the Ocean Carbon Cycle: Why Bigger Was Better," *PLOS ONE* 8, no. 5 (August 26, 2010).

35. Ralph Chami, et al., "Nature's Solution to Climate Change," *Finance & Development* 4, no. 56 (December 2019).

36. James Temple, "Microsoft will invest $1 billion into carbon reduction and removal technologies," *MIT Technology Review*, January 16, 2020.

37. Sarah Frier and Stephen Soper, "Bezos says he's committing $10 billion to fight climate change," Bloomberg Green, February 17, 2020.

CHAPTER 8: TURN THE WORLD VEGAN

1. Dieter Helm, "Agriculture after Brexit," *Oxford Review of Economic Policy* Suppl. 1, no. 33 (March 10, 2017): 124–33; "Major Land Uses," USDA, ers.usda.gov; "Sources of Greenhouse Gas Emissions," EPA, epa.gov; "What is agriculture's share of the overall U.S. economy?," USDA, ers.usda.gov.

2. Brian Machovina, et al., "Biodiversity conservation: the key is reducing meat consumption," *Science of the Total Environment* 536 (December 1, 2015): 419–31.

3. Michael A. Clark, et al., "Global food system emissions could preclude achieving the 1.5° and 2°C climate change targets," *Science* 6517, no. 370 (November 6, 2020): 705–8.

4. David Tilman, et al., "Global food demand and the sustainable intensification of agriculture," *PNAS* 50, no. 108 (October 12, 2011): 20260–64.

5. "Current worldwide annual meat consumption per capita. Food supply—livestock and fish primary equivalent," Food and Agriculture Organization of the United Nations, fao.org.

6. "World Resources Report: Creating a Sustainable Future: A Menu of Solutions to Feed Nearly 10 Billion People by 2050," World Resources Institute, sustainablefoodfuture.org.

7. "Agricultural Policy Monitoring and Evaluation 2018," OECD, June 26, 2018.

8. Magnus Nyström, et al., "Anatomy and resilience of the global production ecosystem," *Nature* 575 (November 6, 2019): 98–108.

9. "Major cuts of greenhouse gas emissions from livestock within reach," Food and Agriculture Organization of the United Nations, fao.org.

10. Carrie Hribar and Mark Schultz, "Understanding Concentrated Animal Feeding Operations and Their Impact on Communities," National Association of Local Boards of Health, 2010.

11. "Global food: waste not, want not," Institute of Mechanical Engineers, November 2, 2013.

12. Helen Harwatt, et al., "Substituting beans for beef as a contribution toward US climate change targets," *Climatic Change* 143 (May 11, 2017): 161–70.

13. J. Poore and T. Nemecek, "Reducing food's environmental impacts through producers and consumers," *Science* 6392, no. 630 (June 1, 2018): 987–92.

14. H. K. Gibbs, et al., "Tropical forests were the primary sources of new agricultural land in the 1980s and 1990s," *PNAS* 38, no. 107 (September 21, 2010): 16732–37.

15. David R. Montgomery, *Growing a Revolution: Bringing Our Soil Back to Life* (W. W. Norton, 2017).

16. Caitlin Hicks Pries, et al., "The whole-soil carbon flux in response to warming," *Science* 6332, no. 335 (March 9, 2017): 1420–23.

17. Robert J. Diaz and Rutger Rosenberg, "Spreading Dead Zones and Consequences for Marine Ecosystems," *Science* 5891, no. 321 (August 15, 2008): 926–29.

18. David Tilman, et al., "Global food demand and the sustainable intensification of agriculture," *PNAS* 50, no. 108 (December 13, 2011): 20260–64.

19. H. C. J. Godfray, et al., "Food Security: The Challenge of Feeding 9 Billion People," *Science* 5967, no. 327 (January 28, 2010): 812–18.

20. C. Nellemann, et al., *The Environmental Food Crisis: The Environment's Role in Averting Future Food Crises* (United Nations Environmental Programme, 2009).

21. "Estimated post-harvest losses of rice in Southeast Asia," Food and Agriculture Organization of the United Nations, fao.org.

22. David Strahan, "Liquid air technologies—a guide to the potential," Liquid Air Energy Network, October 22, 2013.

23. David Tilman, et al., "Saving biodiversity in the era of human-dominated ecosystems," *Biodiversity and Climate Change*, eds. Thomas Lovejoy and Lee Hannah (Yale University Press, 2019).

24. Tim Cashion, et al., "Most fish destined for fishmeal production are food-grade fish," *Fish and Fisheries* 5, no. 18 (February 13, 2017): 837–44.

25. Sergiy Smetana, et al., "Sustainable use of *Hermetia illucens* insect biomass for feed and food: Attributional and consequential life cycle assessment," *Resources, Conservation and Recycling* 144 (May 2019): 285–96.

26. Adeline Mertenat, et al., "Black Soldier Fly biowaste treatment—assessment of global warming potential," *Waste Management* 84 (February 1, 2019): 173–81.

27. Barry M. Popkin, et al., "Global nutrition transition and the pandemic of obesity in developing countries," *Nutrition Reviews* 1, no. 70 (January 2012): 3–21.

28. Stefan M. Pasiakos, et al., "Sources and Amounts of Animal, Dairy, and Plant Protein Intake of US Adults in 2007–2010," *Nutrients* 8, no. 7 (August 21, 2015): 7058–69.

29. Janet Ranganathan, "People Are Eating More Protein than They Need—Especially in Wealthy Regions," World Resources Institute, April 20, 2016.

30. Marco Springmann, et al., "Analysis and valuation of the health and climate change cobenefits of dietary change," *PNAS* 15, no. 113 (April 12, 2016): 4146–51.

31. Marco Springmann, et al., "Health and nutritional aspects of sustainable diet strategies and their association with environmental impacts: a global modelling analysis with country-level detail," *Lancet Planet Health* 10, no. 2 (October 2018): e451–61.

32. Robert D. Kinley, et al., "The red macroalgae *Asparagopsis taxiformis* is a potent natural antimethanogenic that reduces methane production during *in vitro* fermentation with rumen fluid," *Animal Production Science* 3, no. 56 (January 2016): 282–89.

33. Xixi Li, et al., "*Asparagopsis taxiformis* decreases enteric methane production from sheep," *Animal Production Science* 4, no. 58 (August 2016): 681–88.

34. "University of Cambridge: Removing meat 'cut carbon emissions,'" *BBC*, September 10, 2019.

35. "Nutrient Management," Project Drawdown, drawdown .org.

36. S. Sela, et al., "Adapt-N Outperforms Grower-Selected Nitrogen Rates in Northeast and Midwestern United States Strip Trials," *Agronomy Journal* 4, no. 108 (June 17, 2016): 1726–34.

37. Jules Pretty, et al., "Global assessment of agricultural system redesign for sustainable intensification," *Nature Sustainability* 1 (August 14, 2018): 441–46.

38. Zhenling Cui, et al., "Pursuing sustainable productivity with millions of smallholder farmers," *Nature* 555 (March 7, 2018): 363–66.

39. Rattan Lal, "Digging deeper: A holistic perspective of factors affecting soil organic carbon sequestration in agroecosystems," *Global Change Biology* 8, no. 24 (August 2018): 3285–3301.

40. C. P. Reij and E. M. A. Smaling, "Analyzing success in agriculture and land management in Sub-Saharan Africa: Is macro-level gloom obscuring positive micro-level change?" *Land Use Policy* 3, 25 (2008): 410–420; and Gyde Lund and Harvey Kroze, *Africa: Atlas of Our Changing Environment* (United Nations Environment Programme, 2008).

41. "Climate change and food security: risks and responses," Food and Agriculture Organization for the United Nations, fao.org.

42. Chunwu Zhu, et al., "Carbon dioxide (CO_2) levels this century will alter the protein, micronutrients, and vitamin content of rice grains with potential health consequences for the poorest rice-dependent countries," *Science Advances* 5, no. 4 (May 23, 2018).

43. Richard S. Cottrell et al., "Food production shocks across land and sea," *Nature Sustainability* 2, (January 28, 2019): 130–137.

44. "The State of World Fisheries and Aquaculture 2018: Meeting the Sustainable Development Goals," Food and Agriculture Organization of the United Nations, fao.org.

45. Rob Bailey and Laura Wellesley, "Chokepoints and Vulnerabilities in Global Food Trade," Chatham House, June 28, 2017.

46. Delphine Renard and David Tilman, "National food production stabilized by crop diversity," *Nature* 571 (June 19, 2019): 257–60.

47. Elizabeth Dunn, "Scientists Want to Replace Pesticides with Bacteria," *Bloomberg Businessweek*, April 16, 2018.

48. Robert E. Black, et al., "Maternal and child undernutrition: global and regional exposures and health consequences," *The Lancet* 9608, no. 371 (January 19, 2008): 243–60.

49. "Goals of the C4 Rice Project," C4 Rice Project, c4rice.com.

CHAPTER 9: DISCOVER A NEW REALITY

1. M. Markevitch, et al., "Direct constraints on the dark matter self-interaction cross-section from the merging galaxy cluster 1E0657-56," *Astrophysics Journal* 2, no. 606 (January 22, 2004): 819–24.

2. Arthur Loureiro, et al., "Upper Bound of Neutrino Masses from Combined Cosmological Observations and Particle Physics Experiments," *Physical Review Letters* 8, no. 123 (August 23, 2019).

3. Brent Follin, et al., "First Detection of the Acoustic Oscillation Phase Shift Expected from the Cosmic Neutrino Background," *Physical Review Letters* 9, no. 115 (August 28, 2015).

CHAPTER 10: SECOND GENESIS

1. C. Karimkhani, et al., "The global burden of melanoma: results from the Global Burden of Disease Study 2015," *British Journal of Dermatology* 1, no. 177 (July 2017): 134–40.

2. Andre Esteva, et al., "Dermatologist-level classification of skin cancer with deep neural networks," *Nature* 7639, no. 542 (January 25, 2017): 115–18.

3. Huiying Liang, et al., "Evaluation and accurate diagnoses of pediatric diseases using artificial intelligence," *Nature Medicine* 25 (February 11, 2019): 433–38.

4. Hannah Devlin, "AI systems claiming to read emotions pose discrimination risks," *The Guardian*, February 16, 2020.

5. Yilun Wang and Michal Kosinski, "Deep neural networks are more accurate than humans at detecting sexual orientation from facial images," *Journal of Personality and Social Psychology* 2, no. 114 (February 2018): 246–57.

6. Chris Gamble, et al., "Safety-first AI for autonomous data centre cooling and industrial control," DeepMind, August 17, 2018.

7. Hal Hodson, "Revealed: Google AI has access to huge haul of NHS patient data," *New Scientist*, April 29, 2016.

8. Frank Arute, et al., "Quantum supremacy using a programmable superconducting processor," *Nature* 574 (October 23, 2019): 505–10.

9. Kerstin Beer, et al., "Training deep quantum neural networks," *Nature Communications* 808, no. 11 (February 10, 2020).

10. David J. Chalmers, "Facing up to the problem of consciousness," *Journal of Consciousness Studies* 3, no. 2 (1995): 200–19.

11. Jonathan M. Stokes, et al., "A Deep Learning Approach to Antibiotic Discovery," *Cell* 4, no. 180 (February 20, 2020): 688–702.

12. Alex Graves, et al., "Hybrid computing using a neural network with dynamic external memory," *Nature* 538 (October 12, 2016): 471–76.

13. Ray Kurzweil, "Creating Human-level AI: How and When?," February 9, 2017, youtube.com/watch?v=oPyCHwPS04E.

14. Daniel G. Gibson, et al., "Creation of a Bacterial Cell Controlled by a Chemically Synthesized Genome," *Science* 5987, no. 329 (May 20, 2010): 52–56.

15. Charles M. Denby, et al., "Industrial brewing yeast engineered for the production of primary flavor determinants in hopped beer," *Nature Communications* 965, no. 9 (March 20, 2018).

16. Xiaozhou Luo, et al., "Complete biosynthesis of cannabinoids and their unnatural analogues in yeast," *Nature* 7746, no. 567 (February 27, 2019): 123–26.

EPILOGUE: HOW TO SPEND IT

1. Aaron Boley and Michael Byers, "U.S. policy puts the safe development of space at risk," *Science* 6513, no. 370 (October 9, 2020): 174–75.

2. Kate Abnett and Matthew Green, "EU makes world's biggest 'green recovery' pledge—but will it hit the mark?," *Reuters*, July 22, 2020.

3. "General Assembly approves $3 billion UN budget for 2020," UN News, December 27, 2019.

4. Jason Hickel, *Less is More: How Degrowth Will Save the World* (William Heinemann, 2020).

5. Stefan Bringezu, "Possible Target Corridor for Sustainable Use of Global Material Resources," *Resources* 1, no. 4 (March 2015): 25–54.

6. Arundhati Roy, "The pandemic is a portal," *Financial Times*, April 3, 2020.

Acknowledgments

This book began as an idea for a *New Scientist* story before it rapidly ballooned in size. My wonderful agent Patrick Walsh inflated the balloon and steered it to the perfect editor, Mark Ellingham, who has been hugely helpful in shaping the book, feeding me with stories and books and updates, editing with great care, and greatly improving the finished thing. Many thanks to Patrick, what a pleasure to work with you on a second book, and to Mark, and to all at Profile, particularly Nikky Twyman (proofreading), Bill Johncocks (indexing), and Henry Iles (design), who have helped with the project.

Thanks to Emily Wilson, the editor of *New Scientist*, for finding me time to work on this book, and to my colleagues Cat de Lange and Tiffany O'Callaghan. The book has greatly benefited from advice, discussion, and input from dozens of people. At *New Scientist*, Adam Vaughan, Graham Lawton, Jessica Hamzelou, Michael Le Page, Leah Crane, Jacob Aron, Clare Wilson, Valerie Jamieson, Penny Sarchet, Craig Mackie, and many more. The magazine is such an inspiring place to work, and long may it thrive.

One of the great pleasures of being a science writer is getting to talk to people who are among the most knowledgeable in the world for a particular subject. It's a privilege to get your time, and even more so when it is to speculate on what you could do with a vast sum of money that you are never likely to have. So, thanks to all the scientists who generously took time to think and chat about the ideas in this book. I can't list everyone, but I'd like to pick out: Jeremy Farrar, Seth Berkley, Susan Cook-Patton, Mark Jacobson, David Beerling, Dave Reay, Leszek Borysiewicz, Tom Crowther, Rebecca Shaw, David Tilman, David Keith, Peter Wadhams, Kate Larsen, Jim Hansen, Joseph Moore, Jules Pretty, Lynn Dicks, Paul Freemont, Ian Paulsen, Dieter Helm, Emily Nicholson, Johannes Haushofer, Mark Jaccard, Steven Cowley, Charlie Wilson, Cameron Hepburn, Kelly Wanser, Gernot Wagner, Jon Butterworth, Seth Shostak, Julia Steinberger, and Joel Millward-Hopkins.

Several people kindly read drafts of various chapters and fed back helpful comments. Thank you for your time and input: Kate Orkin, Adam Vaughan, Ajay Gambhir, Andy Hector, Sam Krevor, Zeeya Merali, Tom McCauley, and Anil Seth. Any remaining mistakes, of course, are mine.

Many other friends and fellow writers and contacts have provided support and advice in various ways: Gaia Vince, Celeste Biever, Catherine Brahic, Oliver Morton, Victoria James, Simon Aldridge, Cixin Liu (even though I didn't follow your advice on what to do with the money), Kim Stanley Robinson, Christiana Figueres, Tom Rivett-Carnac, Chris Goodall, Tom Belsham, Anit Mukherjee, Chris Faulkes—thanks to you all.

Thanks to my family for your support, especially during lockdown: Ros, John, Steve, Eileen; my mum and Richard, dad and Kathy; Gemma and Phil, Mack and Nate, Scout and Nellie.

My partner Laura Gallagher has influenced and improved the book at every level, from advice on the title all the way through to details on malaria and gene editing and tropical diseases, but also on the whole approach, the tone and feel. Thank you, as ever, for your love, support, and guidance. Finally, our daughters Molly and Iris: I don't have a trillion dollars to spend on you, but this book has been about trying to make the future—your future—better.

Image Credits

Index

Page numbers in *italics* indicate a photograph. Page numbers followed by *n* indicate a footnote.

About the Author

ROWAN HOOPER is a senior editor at *New Scientist* and podcast host of *New Scientist Weekly*. His work has been published in the *The Washington Post, The Wall Street Journal, WIRED*, and *The Economist*. He is also the author of *Superhuman: Life at the Extremes of Our Capacity*. He lives in London.

rowanhooper.contently.com | @rowhoop